KW-020-090

World of Colour
Animal Encyclopedia

Catherine Blow

First published 1980 by
Octopus Books Limited
59 Grosvenor Street
London W1

ISBN 0 7064 1356 3

Original edition © 1978 Editions Chantecler,
a division of Zuidnederlandse Uitgeverij, Aartselaar

Printed in Czechoslovakia
50422

CONTENTS

INTRODUCTION

Some 300 species are mentioned in this book. They are a small but representative selection of the more than one million known species that inhabit the earth. In order to classify these species into a logical order, scientists have devised a system which shows the relationships between one type of animal and another.

The present system of classification is based on the work of the Swedish botanist Carolus Linnaeus who, in the 18th century, published *Systema Naturae*, a book in which each plant and animal known to him was given its own Latin name. Linnaeus differentiated animals mainly by their external appearance. External appearances however can be very deceiving. The ostrich, rhea and emu for example all look very much alike; all are flightless birds with very long necks and a pair of long legs, which inhabit similar habitats and show similar behaviour. They are however members of different species that, although separated by water and vast distances, have evolved to fill the same type of niche and so developed in a similar way. Today the classification of animals is based on physiological, anatomical and evolutionary features together.

The different categories or groups into which animals are classed are known as taxons and the scientific naming of species, taxonomy. The fundamental taxon of one kind of animal is composed of the generic name followed by the specific name. While the specific name may be applied to more than one genus (e.g. the Black Bear, *Ursus americanus* and the yellow-billed Cuckoo *Coccyzus americanus*), the generic name applies to a single genus. Thus, the Black-headed gull is *Larus ridibundus* while the Herring Gull is *Larus argentatus*, the Generic name *Larus* signifying the close relationship between the two species. These gulls share features in common with other species of gull and with the terns, and so are grouped in the family Laridae. The similar anatomical characteristics they have with other families such as waders, including the arrangement of flight feathers and the construction of the voice box, have caused scientists to place them in the order Charadriiformes, the waders, gulls and auks. As they are all warm-blooded, two-legged vertebrates with wings and a feathered body, they are grouped with other orders having these features in the class Aves, the birds. Possessing among other things a notochord, vestiges of gill slits and a dorsal nerve cord, they are placed in the phylum Chordata. To summarize, the main divisions of the animal kingdom, beginning with the species, is genus (plural genera), family, order, class and phylum (plural phyla). However it must be mentioned that further divisions are sometimes necessary, so that the terms superfamily and suborder are frequently encountered. When a species takes a slightly different form from its relatives the term race is applied. For example, there are several different forms of the Two-spotted Ladybird (species *Adalia bipunctata*). This is generally red with two black spots but may also be black with four or six red spots. The difference is reflected in their names, the former being termed *Adalia bipunctata quadrimaculata*, and the latter *Adalia bipunctata sexpustulata*.

As the common, in this case English, name of a species may vary from one region to another (e.g. the European Robin belongs to the genus *Erithacus* and the species called Robin in North America belongs to the genus *Turdus*), the Latin name is necessary for ensuring that one particular kind of animal is being referred to. Each species in this book therefore has the Latin name listed as well as the order, class and family to which it belongs (also the phylum where applicable). It must be noted however that the scientific classification of the animal kingdom is not absolute. In many areas there is disagreement about the families, orders and even classes in which certain animals should be placed. Newly discovered species bring new knowledge to light about the inter-relationships between species. Wherever possible alternative systems of classification have been noted.

Evolution of Animals

Until Charles Darwin published his theory of evolution in 1859 it was generally thought that all life-forms had existed as they were since the beginning of time and that they would remain unchanged until the end of the world. It is now known that both animal and plant life underwent gradual but dramatic changes over the past millions of years, but it is not realized fully that the animals inhabiting the earth today are still undergoing modifications which may result in new forms in the future. In addition, it must be stated that the use of the word 'primitive', applied to many species, does not imply that an animal is inferior to another, nor does the term 'more highly evolved' mean that a species is more capable of surviving in its environment than another. The fact that a species is living is in most cases proof that it has been successful in fulfilling the basic requirements of survival.

Finally, the process of natural evolution fully explained first by Darwin, whereby only the fittest members of a species and the fittest species survive, has been modified over the last centuries by another product of the evolutionary process, man. As the current dominant life-form on the earth, armed with a variety of weapons from guns and snares to pesticides, he has been responsible for the wholesale slaughter and extinction of thousands of animals. Possibly, as the awareness for the need of conservation measures becomes more evident and these are carried out, the numbers of animals we speak about as having existed in the past will not continue to grow until only the human race remains on earth.

MAMMALS

Some 150 million years ago when the dinosaurs dominated the earth, small, furry animals lived inconspicuously among the dense ground vegetation of the forests. Insignificant in terms of numbers and size, they nevertheless already possessed many of the characteristics that eventually would enable them to survive the dinosaurs and take their place as the dominant life-form. These animals had evolved from some form of reptile developing a different body structure which was a completely new adaptation to the environment. One of the first and most important of these developments was that the reptilian scales were gradually replaced with a covering of body hair. Gradually other features were developed, features which are common to the group that scientists now classify as mammals.

The Rise of the Mammals

Reptiles and other cold-blooded animals are at the mercy of the climate. In cold weather they become sluggish as their body temperature drops in response to the external conditions, and in warm weather they have to seek shelter from the sun to prevent their bodies from dehydrating. The development of hair from reptilian scales, like the development of feathers in birds, was the first step in allowing some animals to maintain a constant body temperature, a necessary condition for remaining active regardless of the temperature of the environment. Body heat, produced by the metabolic process of assimilating food, is prevented from escaping by air which is trapped among the hairs.

Equipped with a means of insulating the body against the cold, these early mammal-like creatures also had to be able to stop their body from overheating when the temperature rose. Another characteristic of the class called mammals, the presence of glands in the skin which are distributed over the body, partially solved this problem for many of them. Some of the glands became specialized for releasing water from the body in the form of sweat which, lying on the surface of the skin, cooled it down. (Mammals which lack sweat glands, such as dogs, cool their body by panting.) The hairs, too, could be made to stand erect so that more of the skin was exposed, and during the warm season much of the hair was shed.

Being able to maintain a constant body temperature and thereby becoming more active (being warm-blooded), these animals could live in a variety of different habitats. But as long as the dinosaurs continued to flourish the first mammals held an inferior position, probably leading an insectivorous, nocturnal and secretive existence and competing only with the smaller reptiles that shared the same habitat and feeding habits. It was not until the dinosaurs disappeared, about 64 million years ago, that the opportunity arose for the mammals to diversify and spread out over the earth.

From the time when the first mammals appeared, believed to be 200 million years ago, until the time they began to rise to prominence some 136 million years had elapsed. At the end of this period the mammals were still at a primitive stage of their evolution, but they had developed many of the characteristics found in modern species. Apart from those mentioned above (warm-bloodedness and a body covered with hair), they possessed mammary glands which, like the sweat glands, originated as a specialization of some of the skin glands and secreted a nourishing food for their young. In addition, while some of them retained the egg-laying habits of their ancestors, the majority of them gave birth to live young. These mammals had divided into two main groups—the marsupials, or pouched mammals, and the placental mammals.

Birth and Young

Two types of egg-laying mammals, or monotremes, still survive today in Australia and New Guinea. The Duck-billed Platypus (*Ornithorhynchus anatinus*) and the Echidna or Spiny Anteater and Long-beaked Echidna

(*Tachyglossus aculeatus* and *Zaglossus bruijni*) are very ancient mammals. The males lack external genitals and pass semen through an opening called the cloaca which, during mating, is placed over an identical opening in the female. The cloaca is also present in some other animals, including reptiles and birds. Faeces, urine and sperm, and in the case of the female, eggs, all pass through this single opening. The name monotreme comes from this feature which literally means 'one hole'.

The Platypus female builds a nest at the end of an underground chamber and there lays two tiny soft eggs which she incubates with her warm body. After the young hatch they do not suckle milk from the mother's nipples for she lacks true mammary glands; instead they lap milk from the fur on her belly, secreted there by milk glands that have not developed into teats. The female Echidna feeds her young in the same way and although she also lays eggs these are incubated in a primitive pouch located on the belly. The eggs are carefully deposited there by the mother immediately after being laid.

While the Echidna's pouch is a temporary structure, developing only when required, that of the marsupial is a permanent fold of skin which encases the mammillae or nipples. Like typical mammals, male marsupials possess a penis but in this case it has a forked tip to correspond with the two vaginas of the female. During copulation the sperm passes into each vagina and from there to the double uterus. Marsupial females give birth to live young which are so immature and unlike any other mammal 'babies' that they have been given the name neonates. Before the neonate is born it is nourished by the yolk of the egg. Its birth follows the depletion of this food supply. The neonate then clambers up through the mother's fur, entirely unaided and probably guided only by smell, making its way to the marsupium. Once inside, it clings to the teat, through which milk is pumped in the required quantities for the feeding young, which does not suckle like other mammalian young. There it remains warm and secure for several weeks or, as in the case of the kangaroo, for three months, before it ventures out into the world (see page 16).

For the young of other mammals, such a precarious journey at so early a stage of development is unnecessary owing to the presence of the placenta in the womb. The fertilized egg imbeds itself in the wall of the uterus. At the point where it is attached a mass of tissues develops, the placenta, which passes oxygen and nutrients from the mother's blood stream through the umbilical cord to the embryo (at a later stage called the foetus). Waste products from the embryo are carried in the opposite direction into the mother's blood stream and are expelled through her urine. By this method the foetus is able to remain protected in the womb for a considerable period, its development having proceeded far beyond that of the neonate by the time it is born, so that in most cases it is fully mobile at birth.

Except in the case of precocious young, such as those of the spiny mice (see page 38), birth is followed by a fairly long period of dependency on the mother. The physical dependence while the young is still suckling provides an opportunity for it to learn the techniques of hunting and survival and, in the case of mammals that live in groups, the rules for living within the society. For mammals that live in highly organized groups, such as the chimpanzees and elephants (see pages 26, 106 and 107), infancy is more prolonged than it is for less gregarious mammals. Chimps are totally dependent for at least the first five years of life and elephants for at least the first seven; infancy is followed by a juvenile stage before these mammals reach sexual and social maturity.

Diversity
Compared to other classes, mammals are a fairly small group, with some 4200 species, but they present a great diversity of form and habits. After the decline and ultimate extinction of the dinosaurs, primitive mammals had no serious competitors and were therefore able to spread into a multitude of different habitats. Today mammals are found on the

ground, beneath the ground, in trees, water and even in the air. They vary in size from the tiny Etruscan Shrew (*Suncus etruscus*), weighing 1.5 grams (1/25 of an ounce) to the gigantic Blue Whale, the largest animal that has ever lived, weighing about 150 tonnes. The Blue Whale and other baleen whales live exclusively on a tiny, shrimp-like creature called Krill which floats on the surface of the southern polar seas. A Blue Whale may take up to 4 tonnes of Krill every day, sieving it from the water through huge baleen plates which have, in these whales, taken the place of teeth.

Teeth are important for classifying species and identifying fossils, for they are closely related to the type of food an animal eats and therefore give some indication of its habits. A typical mammal has two sets of teeth during its life, milk teeth and permanent teeth, consisting of incisors, canines, premolars and molars, with a basic number of 44 although there is a great variation in number and type, depending largely on what the animal eats. A few species are completely toothless. The Pangolin feeds only on ants and termites which it gathers with its long, sticky tongue, and has no need for teeth. Toothed whales may have up to 260 teeth (the large flesh-eaters such as the Killer Whale, have many razor-sharp teeth for seizing prey) but the Narwhal has only two, which in the female remain undeveloped. One of the male's grows out of the jaw along a horizontal plane to a length of some 2 metres (7 feet).

Rodents, which comprise the largest group of mammals (Order Rodentia with almost 2000 species), have a pair of very long incisors in both the upper and lower jaws which are used for grinding and gnawing hard foods such as nuts. Covered by enamel only on the front surface, the back wears down faster than the front so that the teeth become chisel-shaped at the tip. The rodent's habit of gnawing at anything solid and hard is caused by the fact that the incisors grow constantly and must be worn down, for otherwise they will grow through the mouth. In this group canine teeth are absent, leaving a wide gap (the diastema, also found in some large grazing mammals such as the horse) between the incisors and the molars.

In carnivores the canines are the most well-developed of the teeth; they are long and extremely sharp for grasping and piercing. Certain premolars and molars, called carnassials, have become specialized for cutting and tearing flesh, while generally the incisors are small and insignificant. In elephants the canines have developed into tusks which, like teeth, become worn with use. The large herbivores, such as the horse and zebra, have sharp incisors for clipping off grasses, while the molars are specialized for grinding the tough vegetation to make it more digestible. These mammals, like man, have a straightforward digestive system not wholly equipped to deal with hard fibres. Others, the ruminants such as sheep, goats and cattle, have complex stomachs and can regurgitate their food and chew it a second time; their stomachs also contain bacteria which help to break down food.

Herbivores do not have to pursue their food, although they may make seasonal migrations in search of fresh pastures. Yet the gathering of food is fraught with danger, especially in the case of grass-eaters which graze on open plains. Plant food is less nutritious than animal food, so herbivores must eat greater amounts than carnivores; this takes more time and makes them more vulnerable to predators. While small mammals, such as squirrels and rabbits, can escape into underground burrows or into trees, the large ungulates (hoofed mammals) must rely on speed. For this purpose they have developed long, strong limbs and hoofed feet for running over hard ground. The fact that most of these animals live in groups is not accidental, for a herd has more eyes, ears and nostrils to sense impending danger. A full-grown, healthy adult is therefore no easy catch for even a Lion or a Cheetah. For this reason hunters usually take sick, injured and young ungulates, or those that have wandered away from the herd and can be taken by surprise.

Carnivores generally capture their prey in

one of three ways: by stalking or ambush or by pursuing them until they drop from exhaustion. The first two methods are usually employed by the family Felidae (cats) whose soft, padded feet aid them in a stealthy approach, and whose strong claws and teeth enable them to make a quick kill. The family Canidae (dogs) use the last method; they have more

Brown Bears eat a large proportion of plant food, and the Panda feeds almost exclusively on bamboo shoots. Although the word carnivore means flesh-eater, other distinguishing characteristics are also used to classify the group, including anatomical features.)

The problem of obtaining a sufficient supply of food has led some mammals to vary

stamina than the felines and can keep up a steady, fast run over a longer distance. As meat contains a large amount of protein, carnivores do not have to eat as often as other animals. Lions, for example, may make only two kills a week and can go for about ten days without feeding; although when food is scarce they will, like many others, sometimes vary their diet and eat smaller prey or even grass. (Other carnivores, such as the Black and

their diet considerably, so that they eat both plants and animals, depending on what is most abundant. Omnivorous animals include rats and mice, and many monkeys. Other mammals, such as the hamster and chipmunk, hoard food in their burrows or in caches around their territory. Some mammals living in cold regions of the world avoid the problem of eating for most of the winter when food is scarce these are the mammals that hibernate.

True hibernating mammals are an exception to the class as a whole in one important feature. Their body temperature varies in the winter with that of the environment, and when it falls below a certain level — in the marmots usually below 15.6 degrees Centigrade (60 degrees Fahrenheit) — their body becomes torpid and they fall into a deep sleep. Broadly speaking, true hibernating species belong to the group of rodents (i.e. marmots and dormice), bats and insectivores (i.e. hedgehogs) although even some of these, such as the hedgehog, will waken at intervals to feed or move about. In the autumn most hibernating mammals increase their intake of food so that they build up reserves in the form of fat. Other mammals that do not hibernate may also store food for energy in this way — the camel, for example, in its hump(s).

An opposite condition occurs in some animals that live in extremely hot climates. During the dry season they burrow beneath the ground and undergo a summer sleep called aestivation; their breathing and the rate of their heart beats become slower as they also do with hibernating animals. The Fat Mice (genus *Steatomys*) of Africa are one example that undergo aestivation.

Species and Orders
The living mammals are generally divided into 19 orders and, of these, 14 orders with 88 species are represented in this book. To give a complete list of the Class Mammalia, the other five orders are: Order Monotremata (the monotremes mentioned above), Order Dermoptera (the flying lemurs), Order Tubulidentata (the Aardvark), Order Hyracoidea (the hyraxes), and Order Sirenia (the seacows). These orders can be said to constitute the minor groups of mammals in the sense that each comprises six or less species and, except for the sirenians, have a restricted geographical distribution.

The relationship between man and his fellow mammals has been, on the whole, a negative one. The number of species endangered or made rare, and in some cases made extinct, through thoughtless and careless actions is astounding. Many mammals have suffered at the hands of man, especially through the clearing of land, which destroys their habitats, and through extensive and often unnecessary hunting. Of the species mentioned in the following pages, those that are now extinct include the Tarpan, the Nubian Ass, Burchell's Zebra, and wild species of Père David's Deer, Przewalski's Horse and the European Bison. And of the 88 species those that are endangered or rare are as follows: all the species of baleen whales, the Red Kangaroo, the Dwarf Chimpanzee, the Giant Armadillo, the Maned Wolf, the Polar Bear, the Otter, the Somali Wild Ass, the White and the Black Rhinoceri, the Wild Boar, the Bush Pig, the Hippopotamus, several species of seal including the Mediterranean Monk Seal, Ross Seal and Ribbon Seal; the Asian Lion, the Brown Hyena, the Tiger, the Jaguar, the Cheetah, two species of llama, the White-tailed Gnu, several species of deer, the Springbok, the Musk Ox, the Moufflon, the Chamois, the Lynx and the African Elephant.

The above list is not a complete one. On the positive side however it should be mentioned that several species have been saved from extinction in recent years. These included the Elephant Seal, American Bison and Père David's Deer which have been returned to the wild from captivity with much success.

KANGAROOS

Kangaroos belong to the primitive group of mammals known as marsupials – animals with abdominal pouches in which the young are nurtured and protected. The family of kangaroos (Macropodidae) comprises some 40 species and includes the tree kangaroos, wallabies and the Red and Great Grey Kangaroos. The last two species are the largest and possibly the best known; they have short forelimbs but long and powerful hind limbs that enable them to jump and hop at considerable speed, with the thick, long tail aiding their balance. The tiny young, known as a neonate, is born at an early stage of its development and immediately makes its way up through the mother's fur to the pouch containing the mammary glands or teats. Here it remains for about six months. Shortly after the first birth the female is again fertilized but the development of the egg is halted until the first-born has abandoned the pouch for good.

Order Marsupialia. Family Macropodidae. Found in Australia and New Guinea, the Red Kangaroo (*Macropus rufus*) grows up to 2 metres (6½ft) tall. The male has a reddish back and grey face with black and white markings on the muzzle, and white throat and underparts. The female is grey on the upperparts and white below. It inhabits open plains. The Great Grey Kangaroo (*M. giganteus*) male grows up to 1.8 metres (6ft) tall, and is mainly greyish. It inhabits eastern and western regions of Australia. Both species are vegetarians and live in large troops. Preyed on mainly by man, they are hunted for sport and for their meat which is used in pet food, and also by farmers because they compete with sheep for pastureland.

EUROPEAN HEDGEHOG

Order Insectivora. Family Erinaceidae, the hedgehogs. European Hedgehog (*Erinaceus europaeus*) grows up to 33.5cm (13in) long, including the tail. Found in Europe and Asia, it inhabits woodland edges, hedgerows and gardens with ample cover. One, sometimes two, litters are produced per year, each with 2–7 young after a gestation period of 30–40 days. Young hedgehogs are born blind, with soft spines. The diet consists of invertebrates, including insects, worms and molluscs, as well as small rodents. It sometimes kills and eats venomous snakes but, contrary to popular belief, it is not immune to poison. It may also take birds' eggs and is reputed to impale apples on its spines and milk cows, the last especially being highly improbable.

Hedgehogs belong to the order Insectivora. Despite their name, they feed on a variety of invertebrates and not only on insects. The European Hedgehog (also known as the Hedgepig) frequents fields, gardens and lowland regions with thick cover. Its most characteristic feature is its prickly coat composed of thousands of spines. The rest of the body, apart from the small naked tail, is covered with fur. By rolling itself up into a tight ball, and thereby concealing its vulnerable areas, the hedgehog protects itself from many predators. It also has a curious habit, known as self-anointing, of covering its body with saliva, the purpose of which is unknown. The hedgehog is nocturnal, hunting between dusk and dawn and resting during the day beneath hedges and shrubbery and in hollows and crevices. It hibernates in a winter nest constructed of leaves and grass, but hibernation may be interrupted.

SHREW

Order Insectivora, Family Sori-
cidae, comprising over 100
members. Common Shrew
(*Sorex araneus*), grows up to
6.2cm (2½in) long with a tail up
to 5.6cm (1½in) long. It often digs
burrows in loose soil but sometimes takes over the holes of
other small animals. It makes a
cup-shaped nest of leaves and
moss. Gestation period lasts
from 13–20 days, and several
litters are produced a year,
each with 5–7 young. Like most
species it is extremely active
day and night with short rest
periods. It does not hibernate,
and must feed almost constantly
as it will die of starvation if
deprived of food for several
hours. Diet consists of insects,
snails, slugs, earthworms and
other invertebrates. The smallest species, the Etruscan Shrew
weighs 1.5g ($\frac{1}{25}$oz).

The large family of shrews is composed of around 20 genera distributed throughout most of the world, although they are absent from Australia and only a few species are found in South America. These tiny, ground-dwelling mammals have thin, pointed snouts, long tails and extremely sharp, pointed teeth. Musk glands on their flanks give them their characteristic, unpleasant odour which makes them distasteful to many would-be predators. The genus *Sorex* predominates in Europe and North America (where they are called Long-tailed Shrews) and includes the Common Shrew of Eurasia. The Pygmy Shrew and Water Shrew of Europe and North America are well-known representatives of other genera. The Common Shrew, like most others, is solitary except in the breeding season and ferociously territorial. The fights that ensue when one member encroaches on another's territory often end in death.

MOLE

Like the shrews to which they are related, true moles are numerous yet seldom seen for they are subterranean mammals, their whole body being highly adapted for living underground. A thick, velvety coat of short, erect fur covers the cylindrical-shaped body. The forelimbs end in broad, five-fingered, clawed hands which bend outwards and are used alternately in a circular motion to shovel earth while the hind feet are braced against the tunnel wall to help the animal push itself forward. The mole's eyes are extremely small and sometimes almost buried in fur, yet they are not useless and probably are able to distinguish at least between light and dark. Both the snout and the short tail have tactile sensory organs and the ears, which lack external pinnae, are protected by a narrow ridge of skin. The mole digs tunnels and chambers which it occupies alone. Although it does not hibernate it stores food in winter, especially earthworms.

Order Insectivora. Family Talpidae – desmans and moles. The genus *Talpa* includes the Eurasian Common Mole (*T. europaea*) growing up to 18cm (7in) long, including the tail. Found in Europe and in northern and central Asia, it inhabits both lowland and upland regions, living in meadows and fields, and broadleaved woods. The female lines the nest chamber with soft materials. After a gestation period of about 4 weeks, a single litter of 3–4 young is born in May or June. Diet consists mainly of earthworms, insect larvae and some molluscs such as slugs. North American species include the unique Starnosed Mole (*Condylura cristata*) which has 22 fleshy tentacles round its nose.

BATS

Order Chiroptera is divided into two main groups: the Greater, or Fruit-eating Bats (suborder Megachiroptera) and the Lesser or Insect-eating Bats (suborder Microchiroptera). The differences between the two are mainly structural rather than dietary, as there is some overlap between the insect-eaters and fruit-eaters. The Fruit-eating Bats are found in the Old World tropics and include flying foxes and fruit bats. Their diet is of fruit such as mangos and bananas but flowers, pollen and insects may also be eaten. The Insect-eating Bats are well-distributed throughout temperate and tropical regions of the Old and New Worlds. Included in this group are the true Vampire or Blood-sucking Bats.

The Order Chiroptera, which means 'hand-wing', contains almost one thousand species of bats, the only mammals that are capable of true flight. Their long forelimbs have exaggerated fingers that act like the ribs on an umbrella, supporting the alar membranes that constitute the wing, or patagium. The wing is actually an extension of the skin on the upper and lower parts of the body. It runs from the arm to the legs and to the tail, if one is present. The thumb is free and has a relatively long claw. Bats fly by reaching upwards and forwards with their arms, cupping the air in their wings as they are brought down and thereby pushing themselves forwards through the air. Although bats are not blind, they use echo-location or sonar to locate food and to navigate around objects in their flight path. By emitting a sound the bat can accurately judge the position and distance of an object from the echo it receives.

PIPISTRELLE

Order Chiroptera. Suborder Microchiroptera. Family Vespertilionidae. Pipistrelle (*Pipistrellus pipistrellus*) grows up to 4.5cm (1.8in) long, with a 25cm (10in) wingspan. Found in Europe, most parts of Asia and in North America, it frequents most kinds of habitats, except high, barren areas, and hibernates in trees and buildings. Females form large nursing colonies in the spring with 300 to 1000 individuals, while the males are solitary or live in small groups during this time. Usually a single young is born after a gestation period of about 44 days. Feeds on small insects, taken and usually eaten in flight, and emerges after sunset to feed. Similar species found in North America include the Eastern and Western Pipistrelles, the smallest in North America.

The genus *Pipistrellus* is composed of around 40 species of small bats found throughout the world, with the exception of South America. The commonest European representative is the Common Bat or Pipistrelle which is also the smallest of the European bats. It is often seen after dusk near houses and is distinguished from other species by its rapid and jerky flight. The upper parts vary from dark orange through medium brown, to dark brown or blackish; the face and ears are dark brown and the under parts pale. The Pipistrelle has short ears but the tragi are relatively large. (The tragus is part of the external ear that grows from the base and is used to focus sound; it is found in most bats, one of several exceptions being the horse-shoe bats.) Like most bats of temperate regions the Pipistrelle mates in the autumn prior to hibernation, but ovulation and fertilization usually do not take place until the spring.

21

LONG-EARED BAT

As their name implies, Long-eared Bats are distinguished by their enormous ears which are almost as long as the body, and are joined together at the inner edge. The genus *Plecotus* comprises two species, both of which are found in Europe and Asia: the Common Long-eared Bat and the Grey Long-eared Bat. The latter has a narrower distribution but otherwise is similar in habits to the Common Long-eared Bat. These bats emerge from their roosts, usually in the crevices of buildings, around dusk; nocturnal activity includes grooming and flying out in search of food. When asleep or in a torpid state they fold their long ears under their wings, but leave the tragi (see page 21) erect. In summer the females form nesting colonies and each usually gives birth to one young. The sexes come together in the autumn and hibernate in groups for the winter. Close relatives in North America are the Big-eared or Lump-nosed Bats (*Corynorhinus*).

Order Chiroptera. Suborder Microchiroptera. Family Vespertilionidae. Common Long-eared Bat (*Plecotus auritus*) grows to 4.8cm (2in) long with ears over 2.8cm (1in) long and a wingspan of 28cm (11in). Found in Europe and Asia, east to Japan, it roosts and hibernates in buildings, in trees and in caves during harsh winters. The flight is slow and fluttering and it is able to hover in the air like a hummingbird while gathering food. It feeds on insects, either taken in flight or gathered from leaves. Grey Long-eared Bat (*P. austriacus*) is slightly larger, and is found in south and central Europe and as far east as the Himalayas.

DOUROUCOULI

Order Primates. Suborder Anthropoidea. Family Cebidae. Douroucouli (*Aotus trivirgatus*) grows to about 30 cm (12 in) long with a tail about 35 cm (14 in) long. Several races are found in Central and South America. It inhabits forest regions and lives in trees from the canopy level almost to the ground. It can be easily tamed and is sometimes kept as a pet. A single young may be born at any time during the year; it clings to the mother's abdominal fur for the first few days but afterwards is also carried by the father. The diet is omnivorous and includes bats, birds, insects, eggs and fruit.

The order Primates contains almost 200 species of mammals, the majority of which are anthropoids, the advanced primates – monkeys, great apes and man. Found in the forests of Central and South America, the Douroucouli is an arboreal creature like all New World monkeys. Exceptional among true monkeys, it is strictly nocturnal, resting during the day in tree hollows and emerging at dusk to hunt. Its nocturnal habits, flattened face and white markings around the eyes have given it the alternative names, Owl Monkey or Night Ape. It travels silently, either along the ground, where it moves about on all fours, or among the trees, leaping from branch to branch, its long tail helping it to balance. It is an extremely noisy vocalist. A dilatable trachea and air sacs in the throat produce the sound which is further amplified by the lips acting like a megaphone.

23

SQUIRREL MONKEY

Four species of Squirrel Monkeys inhabit parts of Central and northern South America where they inhabit the edges of forests bordering rivers. Mainly arboreal and extremely gregarious, Squirrel Monkeys live in large groups of more than 100 individuals, and are never seen in groups of less than 12. Extremely agile climbers and jumpers, they use only their limbs for travelling amongst the trees. Unlike other New World Monkeys, their tails are not prehensile and are used as a balancing aid and not for grasping. Squirrel Monkeys have large skulls and, relative to their body weight, have the largest brain of all primates, including man. Males establish and maintain dominance in the troop by displaying their genitalia. Both males and females regularly rub urine into their fur, especially on the tail, as a means of identification by other members of its species.

Order Primates. Family Cebidae. Common Squirrel Monkey (*Saimiri sciureus*) grows about 30 cm (12 in) long, with a 40 cm (16 in) long tail. Found in Brazil, it inhabits forests bordering rivers. The Red-backed Squirrel Monkey (*S. oerstedii*) is found in Panama. Gestation lasts about 6 months after which the single young is carried by either the father or the mother. Diet includes fruit, nuts, berries, tree frogs, snails, land crabs and butterflies. Although they feed mainly in trees, Squirrel Monkeys will sometimes move down to the ground to look for food, but always in the company of other members of the group.

HAMADRYAS BABOON

Found on both sides of the Red Sea, in north-eastern Africa and Arabia, the Hamadryas or Sacred Baboon was caught and tamed by the Ancient Egyptians, and after death its body was mummified. Like other baboons, they are terrestrial and move about on all four limbs. They gather at night in large groups and during the day disperse into small family groups, usually with one adult male and a harem of up to ten females with their young. The male is a strong disciplinarian and keeps the group close together while it searches for food. Females are smaller than the males, with a brownish coat. The male is grey, with an elaborate mane that covers the shoulders (see right of illustration). The Hamadryas is the fiercest and most aggressive of all the baboons, and will often attack the Leopard, one of its worst enemies, if it feels itself threatened.

Order Primates. Family Cercopithecidae – Old World Monkeys, found in warm regions of the eastern hemisphere. The Hamadryas Baboon (*Papio hamadryas*) grows up to 76 cm (30 in) long, excluding the tail. Found in north-eastern Africa and in parts of Arabia, it inhabits semi-desert steppes. The female mates with the dominant male. Gestation lasts 6 months. Males are fully grown at 7–10 years; females at 4–5 years. They have a life span of about 30–40 years. Diet includes insects, small animals, seeds, roots and grasses.

CHIMPANZEE

Chimpanzees belong to the group of animals known as the apes, mammals that are distinguished from all other primates, except man, by their highly developed brain, the absence of a tail and by their ability to walk in an upright position. The most intelligent members of the family Pongidae, and certainly the most active and inquisitive, Chimpanzees are also extremely gregarious, spending much of their time in groups of 14 to 40 individuals. The bond between mother and baby is stronger than it is between any other members as the juvenile lives with its mother until it is about five years of age. Chimpanzees are arboreal creatures in as much as they sleep in trees and gather much of their food there, yet they rest and play on the ground and can travel quickly on three or four limbs for a considerable distance. They share with man the ability to make and use tools and, as a group, to defend their territory.

Order Primates. Family Pongidae – anthropoid apes – includes the gibbons and the orang utans of Asia and the gorillas of Africa. Chimpanzee (*Pan troglodytes*) males can grow up to 92 cm (37 in) long with a height of 1.7 metres (68 in) when standing. Found in Central Africa, it inhabits forests and the edges of savannahs. Female usually gives birth to one young (rarely, twins). Gestation lasts about 8 months. Diet includes fruit, bark, leaves, shoots and occasionally insects, particularly termites and ants.

SLOTHS

Order Edentata, anteaters, sloths and armadillos. Family Bradypodidae. Two-toed Sloth (of which there are two species of the genus *Choloepus*, distinguished mainly by the number of neck vertebrae – 6 in one species and 7 in the other) grows up to 60 cm (24 in) long. It is found in tropical forests of Brazil, Venezuela, Costa Rica and Colombia and Panama. The diet includes leaves and fruit. Three-toed Sloth (*Bradypus spp*) grows up to 50 cm (20 in) long and is found in tropical forests north of Argentina. Its diet is restricted solely to the leaves of the Cecropia tree, which is found along rivers.

The family of sloths contains only two living genera: the Two-toed Sloth or Unau (illustrated here) and the Three-toed sloth or Ai. Both actually have five toes on the hind feet but differ in the number of fingers on their forelimbs as their names indicate. The fact that these animals spend much of their time hanging immobile from branches has resulted in a number of remarkable adaptations to an upside-down, arboreal life. Their coat, which is thick and shaggy, lies in the opposite direction to that of most other mammals. It grows from the stomach down to the back, and from the wrists and ankles down to the shoulders and rump, so that rain runs smoothly off the body and does not penetrate to the skin. In addition, sloths are hosts to algae which live in grooves in the hairs and turn the coat green (or yellow during droughts) so that the animal is well camouflaged amongst the leaves of sheltering trees.

ARMADILLO

Some twenty species of armadillo are found in South America, with one extending its range north to the southern United States. They all bear bony plates on their bodies which are covered with horn. Two small shields protect the head and tail, two larger ones the shoulders and rump, and between these lie bands of armour which are jointed together but allow the body to bend. The number of bands may vary from 3 to 12 according to the species. Most armadillos burrow in the ground to hunt for food and to protect their vulnerable underparts, although the small Three-banded Armadillo can roll itself tightly into a ball. The Nine-banded Armadillo is remarkable in its reproductive process for the female gives birth to four or more identical young of the same sex, produced by the division of a single fertilized egg.

Order Edentata. Family Dasypodidae, the armadillos. All members have protective shields on their bodies, and in a few the shields are covered with hair. Their diet includes insects and other invertebrates, and small animals which they hunt by burrowing in the ground. Giant Armadillo (*Priodontes gigas*) grows to be almost 1.5 metres (60 in) long and is found in Amazonian forests. Common or Nine-banded Armadillo (*Dasypus novemcinctus*) grows up to 75 cm (30 in) long, and is found from Central America to the southern United States. Three-banded Armadillo (*Tolypeutes tricinctus*) grows up to 40 cm (16 in) long. Found from Guyana to central Argentina, it inhabits open grassland.

PANGOLIN

The Pangolin, or Scaly Anteater, is found only in the Old World, with three species in Asia and four in Africa. Like the Armadillos, its body has a protective cover but in the Pangolin it is composed of horny, overlapping scales that are actually hard, compressed hairs. The chest, belly and throat are naked but are protected when the animal curls up on its side into a ball, the long tail covering the soft underparts. The Pangolin is nocturnal and, as its alternative name suggests, feeds on ants and termites. With its long, clawed feet it can rip open termites' and ants' nests and it then gathers up the insects with its long, sticky tongue. Arboreal species such as the Tree Pangolin have prehensile tails that enable them to hang from tree branches; they gather insects from leaves and from hanging nests.

Order Pholidota – meaning covered with scales – comprises one family, Manidae. All Pangolins are completely toothless, and their tongues are over 30 cm (12 in) long. Giant Pangolin (*Manis gigantea*) grows up to 150 cm (60 in) long. Found in tropical Africa, it digs burrows and is active at night. Cape Pangolin (*Manis temmincki*) grows up to 50 cm (20 in) long, and its tail can reach to 35 cm (14 in) long. Found in South and East Africa, it is mainly a ground-dweller but will occasionally climb trees. Asian species include the Chinese Pangolin (*Manis pentadactyla*) of southern China, Nepal and Indonesia, and the Indian Pangolin (*M. crassicaudata*) of India and Sri Lanka.

RABBIT

Some 70 species of rabbits and hares make up the family Leporidae, with representatives in most parts of the world. The European Wild Rabbit, originally restricted to north-west Africa and the Iberian Peninsula, has spread over the past thousand years to most of western Europe, and has been introduced into other areas, including the United States, New Zealand, Chile and Australia. The coat is predominantly yellowish-brown and the ears short without markings; the white undertail is conspicuous when the animal runs. Extremely prolific and a voracious eater of all types of vegetation, the Wild Rabbit is a serious agricultural pest where its numbers are not controlled by predators or by the disease known as myxomatosis. Unlike hares, rabbits are gregarious and excavate a complex of burrows which may house a hundred or so individuals. They are timid and nervous and usually only emerge from their warrens at dusk.

Order Lagomorpha – rabbits hares and pikas. Family Leporidae. European Wild or Common Rabbit (*Oryctolagus cuniculus*) grows up to 40cm (16 in) long, excluding the tail. It inhabits open country in both lowlands and hills, usually where the soil is sandy. Females become fertile at about 6 months and the gestation period lasts some 30 days. There are usually 7 litters per year, each with 3–7 young. The nest is made in the warren or in separate small burrows (known as stops). It is constructed of dried grasses and lined with down from the doe's breast. The diet includes agricultural crops and bark from trees. Close North American relatives are the rabbit-like cottontails (*Sylvilagus spp*).

HARE

Order Lagomorpha. Family Leporidae. Brown Hare (*Lepus capenis*, sometimes classed as *L. europaeus*) grows up to 71cm (28in) long, including the tail. Native to Europe, Asia and Africa, it is now also found in North America, Chile, Australia and New Zealand. It inhabits open country such as moors, fields and pastures, and occasionally woodland. Gestation lasts 42 days, and up to 4 litters are produced a year, each with 1–5 young. Leverets are born in a shallow depression or 'form'. The vegetarian diet includes grass, root crops and tree bark, but the Hare is not as destructive as rabbits. Mountain Hare (*L. timidus*); Snowshoe Rabbit (*L. americanus*).

Except in the breeding season, hares are solitary creatures. They inhabit open grassland in most parts of the world. Generally they are larger than rabbits, with long ears and large hind feet and eyes. Native to Europe, Asia and Africa, they are now found on most continents, having been introduced into North and South America, Australia and New Zealand. The Hare has a brown coat with white underparts and cheeks, black tips to the ears and a broad black stripe on the white tail. In winter the coat becomes thick and reddish. Although hares are mainly nocturnal and feed after dusk they may also be seen during the day, bounding rapidly away when danger threatens or in some cases flattening themselves against the ground with their ears laid back against their body. The genus *Lepus* also includes the Mountain or Blue Hare, and the Jack Rabbit and Snowshoe Rabbit which, despite their names, are in fact hares.

ALPINE MARMOT

Order Rodentia. Suborder Sciuromorpha – squirrels and their allies. Family Sciuridae, comprises squirrels, prairie dogs and marmots. Alpine Marmot (*Marmota marmota*) has a body length of 56cm (22in), with a tail up to 20cm (8in) long. It lives in the Alps, Pyrenees, Tatras, Carpathians and in the Black Forest. Diet consists of green vegetation. Gestation lasts from 35–42 days producing a litter of 2 to 6 young. Its predators include birds of prey, especially the Imperial Eagle. Related North American species include the Ground Hog or Woodchuck (*M. monax*) which inhabits woodland and farms.

The Alpine Marmot is native to the Alps but has been introduced into other mountainous regions of Europe. Related species are found in North America and Asia. Like most rodents, it is a burrowing mammal and makes extensive underground galleries and chambers which in the winter are used for hibernation. During spring and summer the Alpine Marmot gathers in groups of about 12 individuals and spends much of its time above ground basking in the sun, feeding and grooming. Sentries posted about the colonies give a shrill whistle when intruders approach and the members flee rapidly underground until the danger has passed. By autumn they become fat on vegetation gathered in the proximity of their burrows, and prepare their winter quarters by lining the chambers with dried grass. Hibernation generally lasts from the end of September until late April, when they live off their reserves of fat.

RED SQUIRREL

Found throughout wooded areas of Europe and Asia, the Red Squirrel is an arboreal rodent that is an adept climber and jumper. On the ground it moves by making short jumps or by running with its long tail held out behind, often pausing on its hind legs to sniff the air. The Red Squirrel usually makes its drey next to the trunk of a tree, but it sometimes builds in tree hollows. In both cases it constructs the nest from twigs and leaves and lines it with soft materials such as grass and moss. This rodent is diurnal and is active throughout the year, although during adverse weather it rests in its drey. Small stores of food are made in various locations round the territory, in the drey, on the ground and in tree hollows. Close relatives, the North American Grey Squirrels, are widespread throughout North America; one species has been introduced into England and Wales where it is now abundant in wooded areas and parks.

Order Rodentia. Suborder Sciuromorpha. Family Sciuridae. Red Squirrel (*Sciurus vulgaris*) grows to 22 cm (8½ in) long, with a tail about 18 cm (7 in) long. Found in Europe and Asia, it inhabits large, dense coniferous forests, mixed woods and parks, and occasionally broad-leaved woods. Gestation lasts from 36–42 days, and 2 litters are produced a year, each of 3–8 young. Diet includes seeds, nuts, leaves, fungi, some animal food such as insects and eggs. Grey Squirrel (*S. carolinensis*) found in eastern North America as far west as the prairies, was introduced into England and Wales between the 1870s and 1920s. Contrary to popular belief it has not caused a decline in the Red Squirrel population.

Order Rodentia. Suborder Sciuromorpha. Family Sciuridae. Eastern Chipmunk (*Tamias striatus*) grows to 15cm (6in) long, with a tail of 10cm (4in). It is found in pine and birch forests in eastern Canada and the eastern United States. Gestation lasts about 35 days, and usually one litter is produced a year with about 6 young. Diet includes nuts, berries, and maize. Western Chipmunk (*Eutamias spp*), grows to 9–15cm (3½–6in) with a tail 6.5–13cm (2½–5in) long. Five species are found in western Canada and the western United States. They include the Alpine Chipmunk, (*E. alpinus*), the Yellowline Chipmunk (*E. amoenus*) and the Colorado Chipmunk (*E. quadrivittatus*).

CHIPMUNK

Chipmunks belong to the group of squirrels and, with the exception of one Siberian species, are exclusive to North America. Although they are excellent climbers they are predominantly terrestrial rodents. They excavate burrows in the earth in which they build their nests and hibernate during the winter. Chipmunks inhabit wooded areas and feed on a variety of nuts which they carry in their cheek pouches and store in large quantities for use during the winter, for they do not undergo deep, continuous sleep but awaken at intervals to feed. Two genera of chipmunks are distinguished: *Eutamias*, occurring in western North America and Asia, and *Tamias*, comprising one species found only in eastern North America. The latter is the largest species and has less distinct stripes on its back than the former.

GOLDEN HAMSTER

Order Rodentia. Suborder Myomorpha, Family Muridae. Golden Hamster (*Mesocricetus auratus*) grows up to 18cm (7in) long, with a tail up to 1.2cm (½in) long. It was originally from Syria. Gestation lasts about 16 days, with 5 to 6 young being produced. They are solitary animals which should only be placed together for mating, as the female is aggressive. Diet includes grains, cereals and green vegetables. Common Hamster (*Cricetus cricetus*) grows up to 34cm (13½in) long, and the tail to 6.5cm (2.6in) long. Found in eastern Europe and western Asia, it inhabits grasslands. The diet includes seeds and grains.

Although the Common Hamster is found in the wild in eastern Europe and western Asia, all the Golden Hamsters known today are kept either as pets or as laboratory animals. The entire stock are the descendants of one female and her three young which were discovered in Syria in 1930 and then brought to Europe. (In all, there were twelve young originally but only three of them survived.) Curiously, there are no records of these rodents having been seen in the wild before or since. Like the Common Hamster to which they are related, Golden Hamsters are essentially nocturnal and may hibernate for several weeks during the winter. Both species carry food in cheek pouches which they deposit as a food store, and like other rodents gnaw at hard objects to wear away their constantly-growing incisors.

VOLE

Voles belong to a large group of mouse-like rodents (sub-order Myomorpha) which also includes mice and rats. Very prolific, in periods when food is abundant their population increases to epidemic proportions (periods known as vole years), causing enormous damage to crops. Voles differ from true rats and mice by their shorter tails, small ears and eyes and their long, rough coats. Their noses are blunt and unlike most true mice they never live in close association with man. Two main genera represented in Europe and North America are *Microtus*, sometimes called the field voles, and *Clethrionomys* or red-backed voles. The Field Vole of western Europe is a common inhabitant of rough grassland, but it is sometimes found in woodlands and hedges. It has a greyish-brown coat and a short tail (never more than half its body length) and like many voles it is active during the day and night.

Order Rodentia. Family Cricetidae: Field Vole (*Microtus arvalis*) grows up to 11.5cm (4½in) long, with a tail up to 4.6cm (1.8in) long. It is found in western Europe, in grassland, woodland and hedges. It breeds from spring to late autumn, each litter producing 4–6 young. Diet includes grass stems and their leaves. A close North American relative, the Meadow Vole (*M. pennsylvanicus*), is found in grassy and boggy areas of northern United States and Canada. Bank Vole (*Clethrionomys glareolus*), is found throughout most of Europe and inhabits woodland and scrub. It may be up to 11cm (4.4in) long, with the tail up to 5.5cm (2.2in) long. Diet includes fleshy fruits and seeds.

MUSKRAT

Order Rodentia. Family Cricetidae. Muskrat (*Ondatra zibeth-
ica*) grows to 40 cm (16 in) long.
Native to the United States and
Canada, it is now also found in
central Europe and the USSR. It
inhabits still and running waters
such as ponds and rivers and, in
Europe, also canals. It has 3–4
litters a year, each with 4–6
young. Diet includes aquatic
vegetation and sometimes
mussels, crayfish and other
water animals. Its predators include weasels and man. A close
relative, the Florida Water Rat
(*Neofiber alleni*) is found in
swamps in the south-eastern
United States, from eastern
Florida to Georgia.

Closely related to the vole, the Muskrat is native to
North America where it leads a largely aquatic life in
ponds and rivers. A fast and powerful swimmer, the
Muskrat propels itself through the water with its strong
limbs and long tail, which is flattened towards the end.
Where there are steep banks along the waterways it
excavates a burrow with an entrance above the water
level and one below. Like the beaver it will also construct
a mounded lodge on the water of reeds, sticks and other
plant material. This rodent has a thick, soft coat, brown
above and grey below. Musk glands are present in the
males, giving them their characteristic strong odour. The
Muskrat's pelt is highly valued by the fur industry and
the musk glands are used in the manufacture of perfume.
Introduced into Bohemia in the early 1900s, this rodent
is now widespread throughout much of central Europe
where it is a pest owing to its habit of burrowing.

SPINY MOUSE

The Cairo or Egyptian Spiny Mouse illustrated here is one of several species found from India south-west into most of Africa. The spiny mice are distinguished by the sharp prickles or spines on their backs that have evolved, like those of the Porcupine, from soft hairs, and which afford them some protection against predators. The Egyptian Spiny Mouse is a diurnal rodent of sandy or rocky regions, living in small groups and taking refuge amongst rock crevices. Its coat varies from grey to a sandy colour, depending on its surroundings, and the whiskers are long and stiff. The tiny young become fully active immediately after birth and are born with their eyes open; after some two weeks they are independent and sexually mature. Two closely related species are the Golden Spiny Mouse, which has a similar range, and the Cape Spiny Mouse which is restricted to South Africa.

Order Rodentia. Family Muridae – Old World rats and mice. Spiny mice are found in Asia and Africa and inhabit dry, sandy or rocky regions. Gestation lasts about 38 days and there are 2–3 young in each litter. The birth of fully active, precocious young is unique among the Murids. Egyptian Spiny Mouse (*Acomys cahirinus*) grows to about 9.5–12.8cm (3½–5in) long. Diet includes insects and leaves. It is preyed on mainly by snakes. Cape Spiny Mouse (*A. subspinosus*), has a less prickly coat. The Golden Spiny Mouse (*A. russatus*) differs by having black soles on the feet. New World Spiny Mouse, the South American Spiny Mouse, belongs to the family Cricetidae.

HOUSE MOUSE

The genus *Mus* contains many species of true mice found mainly in eastern Asia and Africa although one, now known as the House Mouse, is a native of the European and Asiatic steppes which with the help of man, has spread to all parts of the world. Found in both urban and rural communities, it generally lives in close association with man, inhabiting his buildings and taking his food throughout the year, although rural forms may spend the summer months in fields feeding on insects, seeds and grains. Extremely adaptable, House Mice can live under an amazing variety of conditions, even tolerating total darkness and low temperatures. They are agile climbers capable of scaling walls and cables, good runners and jumpers, and adequate swimmers. Mainly nocturnal, they can be detected by their high-pitched squeak, droppings, footprints and the damage they do by gnawing.

Order Rodentia. Suborder Myomorpha. Family Muridae – Old World rats and mice. House Mouse (*Mus musculus*) grows to 9 cm ($3\frac{1}{2}$ in) long, excluding the tail. It is distributed world-wide, inhabiting buildings and sometimes open country. Gestation lasts about 20 days, and litters of 4 to 8 young are produced throughout the year by commensal (those associating with man) populations. Diet includes human food and refuse, as well as seeds, grains and some insects; occasionally an odd choice of domestic articles such as soap are also taken. Colouring is generally greyish brown, and sometimes with paler underparts. The tail is long and relatively thick with obvious scales.

LONG-TAILED FIELD MOUSE

Order Rodentia. Suborder Myomorpha. Family Muridae. Long-tailed Field Mouse (*Apodemus sylvaticus*) can be up to 11cm (4½in) long, with a tail up to 10cm (4in) long. Found in Europe, Asia, North Africa and Iceland, it inhabits woodlands (both coniferous and broad-leaved), gardens, hedgerows and fields with some cover. Very occasionally it is found in houses, but more often in farm outbuildings. Diet includes seeds, buds, fruit, nuts and arthropods. Yellow-necked Field Mouse (*A. flavicollis*) grows up to 11.5cm (4½in) long, with a tail up to 12.7cm (5in) long. It is found in England, Wales and continental Europe, with the exception of Spain, Italy and the Benelux countries.

The commonest group of mice found in Europe belong to the genus *Apodemus*, which includes the Long-tailed Field Mouse or Wood Mouse. This mouse lives in a variety of habitats, from fields and woods to gardens and hedgerows. Small groups, headed by a dominant male, may have a territorial range of up to 2.4 hectares (6 acres) in which they breed and forage for food. For most of the year these mice are extremely active and, like the House Mouse, are agile runners and jumpers and excellent climbers. In winter they become somewhat sluggish and feed mainly on food they have hoarded during the autumn. Sometimes confused with the Yellow-necked Mouse, another woodland species, the Long-tailed Field Mouse has a greyish-brown back, paler underparts and frequently a yellowish streak along its flanks. The Yellow-necked Mouse has a longer tail, usually longer than its body length, and is yellow.

HARVEST MOUSE

This tiny rodent is an inhabitant of damp and wet regions with tall, dense vegetation, and is found in many parts of Europe and in Asia. It has a reddish-brown back and head and white underparts; the nose is blunt and the ears small and hairy. Extremely fragile and light, it climbs amongst grasses, reeds and cereal crops, grasping their stems with the prehensile tip of its long tail. In the breeding season the Harvest Mouse builds a spherical nest some 60 centimetres (24 inches) above the ground. The base is formed of long leaves bent over to bridge the stems to which they are attached. Grass leaves are then woven together to form the body of the nest, which is approximately 10 centimetres (4 inches) in diameter. Prior to winter the Harvest Mouse abandons its summer habitat. In dry areas it moves along runways on the ground but in wet, marshy regions it moves to drier areas and many are found in hayricks.

Order Rodentia. Suborder Myomorpha. Family Muridae. Harvest Mouse (*Micromys minutus*) grows to 10 cm (4 in) long, and has a tail of about equal length. It is found in England, Wales, most parts of continental Europe except Spain, and in Asia as far as Japan. It inhabits damp meadows, fields, reed beds, ditches and grassy hedgerows. Gestation lasts from 17 to 19 days, and 2 to 3 litters are produced per year, each with 3 to 7 young. The nest is cylindrical with an opening at the side, and the chamber is lined with short grass stems. Diet includes seeds, grains, fruits, berries and insects.

Order Rodentia. Suborder Myomorpha. Family Gliridae. Common Dormouse (*Muscardinus avellanarius*) grows up to 8.5cm (3½in) long, with a tail about 6.8cm (2½in) long. Gestation lasts about 22–24 days with usually 1–2 litters a year, each with about 4 young. Found in most parts of Europe, but absent from Spain, Denmark and northern Scandinavia, it inhabits broadleaved woods, especially those with sweet chestnut, hazel and birch trees. It is also found in hedgerows and copses. Diet includes fruit, nuts and berries, and possibly insects. Fat Dormouse (*Glis glis*) can be up to 17.5cm (7in) long, with a tail up to 15cm (6in) long. It is most common in southern and eastern Europe and is found in deciduous woods, orchards and gardens.

COMMON DORMOUSE

Seven genera of dormice make up the Old World family Gliridae, most of which are found in Europe and Asia, with one genus in Africa. Unlike the murids, dormice hibernate during the winter and many of them are arboreal. One of the best-known is the Common Dormouse, also known as the Hazelmouse and the Dory Mouse. An agile climber, for much of the year it lives above ground, often quite high up among the branches of shrubs and trees where it builds a round nest of leaves and mosses. This dormouse is orange-brown on the upper parts and pale buff underneath; the tail is covered with short hair. In the autumn its weight increases because of fat reserves and it makes a winter nest on the ground or below the surface, where it will remain from October until April. The Common Dormouse and its relative, the Fat or Edible Dormouse, are both nocturnal creatures.

Order Rodentia. Family Gliridae. Garden Dormouse (*Eliomys quercinus*) grows to 15cm (6in) long, and has a tail up to 12.5cm (5in) long. Found in Europe, central Asia, and North Africa, it inhabits both coniferous and broadleaved woods, vineyards, orchards, large gardens and favours hilly or mountainous country. The breeding nest is made of moss and leaves and is lined with soft materials. It is usually made in tree holes, rock crevices, or bushes, and sometimes in nesting boxes or abandoned birds' nests and, rarely, in squirrel dreys. Gestation lasts about 23 days and one litter is produced per year, with 3–7 young. Diet includes fruit, berries, seeds, snails, eggs, young birds, small mammals, and insects. It hibernates during the winter.

GARDEN DORMOUSE

Larger than the Common Dormouse, the Garden Dormouse is also found in woodland, especially broad-leaved woods. However, it is more common than its relative in orchards, vineyards and gardens, and lives at quite high altitudes in mountainous regions. This dormouse has a greyish-brown coat with white underparts and the broad tail has a white and black tuft at the tip. The face is marked with black eye-stripes. Its hearing is extremely good and is used to locate prey, for the Garden Dormouse is more of a hunter than the Common Dormouse, taking young birds and mammals smaller than itself as well as insects. Although it is not very sociable in company with others, especially during the breeding season, this dormouse is very noisy, emitting whistling, snoring and growling sounds. Like many species it is an able climber and often nests and feeds in trees and branches.

GUINEA PIG

Order Rodentia. Suborder Hystricomorpha. Family Caviidae. Brazilian or Peruvian Cavy (*Cavia aperea*), found in South America. Domesticated Guinea Pig (*C. aperea porcellus*) grows to about 20–30 cm (8–12 in) long. Wild species inhabit a variety of places with thick cover. Gestation lasts about 65 days, and there may be 2 or 3 litters a year, each with 2–6 young. The Brazilian or Peruvian Cavy is similar to the domestic form but is more slender. It may live for 6 to 8 years and feeds in captivity on oats, green vegetables, hay and bran.

The Guinea Pig belongs to the Family Caviidae, the cavies, a New World group of South America. One of the best-known members, the domesticated Guinea Pig, was introduced into Europe in the 16th century. Its association with man goes back further than this, for the Incas had tamed them, eaten them and used them as sacrificial offerings to their gods. Sometimes called Indian Pigs or Indian Rabbits, their more popular name possibly arose from the fact that the earliest imports arrived on slave ships which sailed from Guinea in Africa. Today their descendants are found in all parts of the world and are used as laboratory animals or kept as pets. The races range from smooth short-haired to long-haired Angoras, all of which come in a variety of colours. The wild form (the Brazilian or Peruvian Cavy) lives in small communities, making long runways in the grass or other vegetation.

MARA

The Mara, or Patagonian Cavy, is found exclusively in South America where it inhabits the open, dry grassy plains of southern Argentina and Patagonia. A large rodent, it rather resembles a hare, especially in its manner of fleeing from predators. It runs on its long legs in a series of bounds, often out-distancing its pursuers and leaping over small bushes and shrubs. The hind feet are padded and have three clawed toes. The front feet have four toes, also with sharp claws. The Mara excavates large burrows in hill slopes, although it sometimes takes over the burrows of other animals. The burrows are used for sleeping and nesting but the Mara will also rest in long grass. It is a diurnal animal, feeding during the day, and has long eyelashes to protect its eyes from the sun's rays.

Order Rodentia. Suborder Hystricomorpha. Family Caviidae, the cavies. Mara (*Dolichotis patagonum*) is about 75 cm (30 in) long and is found in Patagonia and on the pampas of Argentina. The Mara has a pale coat and white patch on its rump which is most obvious when the animal is running. It is used as a signal of danger. Two litters are produced a year, each with 2–5 young. Diet includes roots, grasses and plant stems. A gregarious animal, it lives in groups of 10–30 individuals. A similar species, called the Salt Desert Cavy (*Pediolagus salinicola*) is smaller and lacks the white rump; it is found in the salt deserts of Argentina.

COYPU

Like the Mara (page 45), the Coypu is a South American rodent but it has a much wider distribution in its native area, being found from Peru south to the tip of the continent. During the last century it was introduced into several countries including the United States, England and continental Europe. The Coypu is valued and hunted both for its soft fur undercoat (known as nutria and used by the fur industry) and for its meat. This activity has produced a decline in the number of coypus in parts of South America. In Europe it is often considered a pest; in England for example it has caused damage to dykes, crops and natural vegetation. The Coypu's fine grey undercoat is protected by a coarse outer coat of glossy brownish-yellow guard hairs. It has strong hind and forelimbs, each with five toes or fingers, which aid it in swimming. When threatened this animal will run with a galloping movement or dive in water.

Order Rodentia. Suborder Hystricomorpha. Family Capromyidae, composed of one species. Coypu (*Myocastor coypus*) has an average body length of 46cm (18in). Native to South America, it is now found in England, parts of continental Europe, Israel, the USSR, Kenya and the United States. Gestation lasts some 130 days and there are usually 5 young in each litter. The young are precocious and are able to feed on solids and swim within a day or two of their birth. Diet consists mainly of aquatic vegetation.

SEALIONS

Order Pinnipedia. Family Otariidae, composed of fur seals and sealions. California Sealion (*Zalophus californianus*) males grow up to 1.8 metres (6ft) long. They are found along the coast of California, in the Galapagos Islands and south of Japan. Males are polygamous and gather harems of up to 40 cows. Gestation lasts almost 12 months, after which a single pup is born which can swim at about 2 weeks. Diet consists of squid, octopus and sometimes fish. Like other sealions they also swallow hard objects such as stones, the purpose of which is unknown. South American Sealion or Southern Sealion (*Otaria byronia*) males grow up to 2.2 metres (7½ft) long. Most numerous of sealions, they are found off South America.

The group of marine mammals known as the Pinnipedia is composed of the true seals, walruses and eared seals. Sealions belong to the last group. As their name indicates, the eared seals have a small external ear (absent in most true seals) and are further distinguished from true seals by their longer necks and their flexible limbs which enable them to walk on all-fours. The bulls of the Southern and Steller's Sealions grow long manes on their necks, which has resulted in the allusion to the lion in their name. Like most other members of the Pinnipedia, sealions congregate in large numbers to breed, coming ashore once a year for this purpose. This habit has made them extremely vulnerable to predation by man. Many of the bulls maintain large territories and are polygamous; Steller's Sealion bulls may have up to 80 cows in a harem. Some species engage in vicious battles to defend their territories.

SEALS

Most of the phocids, or true seals, inhabit the cold waters of the southern and northern hemispheres. A dense fur coat and a thick layer of blubber beneath the skin enable them to withstand extremely low temperatures. Almost all of these seals lack an external ear. There is little definition between the head and body, and they move in an entirely different way to the eared seals, both in the water and on land. Sealions and fur seals propel themselves in the water mainly with their forelimbs, but the true seals swim with their hind flippers held soles together; they push their body forwards by moving the tail-end of the body from side to side. This means of propulsion and the smooth, streamlined body make these seals graceful and fast swimmers. Yet on land they are clumsy and have to hump themselves along with their front limbs, the hind ones trailing behind or held above the ground.

Order Pinnipedia. Family Phocidae, the true seals, composed of 4 subfamilies. Common or Harbour Seal (*Phoca vitulina*) males grow to 1.8 metres (6ft) long. Found in Northern Atlantic and Pacific Oceans and in Seal and Harrison Lakes in Quebec, they inhabit shallow waters. These seals spend much of their time hauled up on sand banks, mud banks and rocks, usually in small groups of 2 to 40 individuals. A single pup is born in the summer, sometimes in the water. Diet consists of fish. Southern species include the Crabeater (*Lobodon carcinophagus*) and Leopard Seal (*Hydrurga leptonyx*) found around the Antarctic. Both travel as far as Australia and New Zealand.

ELEPHANT SEAL

The Elephant Seal, of which there are two species, is the largest of the Pinnipedia, the males being several times larger than the walruses. They are true seals which belong to the sub-family Cystophorinae, the proboscis seals. The males have a long nose which they inflate during the breeding season to a length of about 60 centimetres (24 inches). The inflated proboscis acts as a resonator, amplifying the sound of the bulls' roars which accompany threatening behaviour. Fights amongst the bulls are ferocious and bloody, many of them sustaining large gaping wounds. The females give birth about a week after coming ashore, to young conceived the previous year. The birth is soon followed by mating. Despite their size and weight of around 1000 kg (2200 lb), these seals are adept swimmers and divers, submerging to hundreds of metres in search of fish. They come ashore in September to breed and in January to moult.

Order Pinnipedia. Family Phocidae. Northern Elephant Seal (*Mirounga angustirostris*) males grow up to 4.8 metres (16ft) long. Found off the coast of Southern California and Baja California. Southern Elephant Seal (*M. leonina*) males grow up to 6 metres (20ft) long. They are found in waters off the Antarctic mainland and in those off the South American coast. Males may gather harems of up to 50 cows. Gestation lasts about 12 months, and the single pup weighs about 36kg (80lb) at birth, increasing by some 9kg (20lb) per day. Diet consists of cephalopods and fish. Elephant Seals were slaughtered in huge numbers at the beginning of the 19th Century and became almost extinct. They are now a protected species.

LION

Like most of the Felidae, the Lion is superbly adapted for hunting. Whereas other members lead a generally solitary existence, lions live in groups, usually consisting of one or two males, several females and their cubs. Although the male is reputed to be the most superior of all animals, it is in fact the lioness who hunts, the male securing and maintaining the territory against intruders so that the pride can function as a cohesive unit. Lacking the stamina of dogs, a lioness hunting alone will ambush her prey, making a short, fast sprint before bringing it down. Several lionesses hunting together will stalk a herd, encircle it and then get the animals to stampede towards one or two of the group who will then make the kill. The male is the first to feed on the carcass and in times when prey is scarce, or the kill is small, the females and cubs may go hungry. It has recently been observed that lions will sometimes steal the kill of other animals.

Order Carnivora. Family Felidae. Lion (*Panthera leo*); males grow up to 2.8 metres (9¼ft) long, excluding the tail which may measure 85 cm (35 in) in length. Originally found in Europe (cave paintings testify to their existence as far west as Spain), most of Africa and Asia, from Turkey to India, today they are restricted to the African savannahs and the Gir Forest in India. Gestation lasts 112 days and produces 2 to 3 cubs. Male cubs begin to grow manes at about two years, although not all males have manes. Diet consists mainly of gnus, antelopes and zebras. They are preyed on only by man.

TIGER

Order Carnivora. Family Felidae. Tiger (*Panthera tigris*); males grow up to 3.9 metres (13 ft) long including the tail. They are found from the USSR and China in the north, south to India, Sumatra, Java and Bali. Diet consists mainly of large hoofed mammals such as the buffalo, but small animals such as wild boar and peacocks are also taken. Near villages, domestic cattle are hunted and, in some instances, man. Gestation lasts about 3 months with 1–2 young produced each year. Solitary except during the breeding season, one tiger may occupy an area of 650 square kilometres (250 square miles).

Although they differ in habits and outward appearance, tigers are closely related to lions and in fact belong to the same genus, *Panthera*. Originally confined to the northern forests of Asia, they spread as far south as the Equator on the continent and to the islands of Java, Sumatra and Bali. Today their numbers are declining drastically and they are in danger of extinction. These carnivores live in open areas fringed by forests, usually resting under cover during the day and emerging at dusk to hunt. Like the lions, tigers lack stamina and rely on stalking and ambush to bring down their prey. They attack from the rear, making their kill by strangulation. Tigers are easily distinguished from other members of the Felidae by their strikingly marked coats, but the patterning and boldness of the pelage varies from race to race, as does their size, although tigers are generally larger than lions.

LEOPARD

Order Carnivora. Family Felidae. Leopard (*Panthera pardus*) grows to 150cm (60in) long, with a tail 90cm (36in) long. It is found in Africa, and Asia, including China, parts of the Soviet Union and India. Males and females associate only briefly for mating which may take place at any time of the year. Gestation lasts about 100 days, with 1–3 and sometimes as many as 6 young being produced. Diet is extremely varied and includes monkeys, antelopes, wild dogs, fish, domestic animals, and occasionally carrion. It is hunted widely for its skin. Jaguar (*P. onca*), found in the southern United States, Central and South America, inhabits forests and open plains.

One of the most adaptable and versatile of the Felidae, the Leopard is a solitary creature outside the breeding season; the only groups that are seen together consist of a female and her young. Found in Africa and Asia, it ranges north to China and parts of the Soviet Union, and south as far as South Africa. Although it is predominantly a creature of the forest it is also found in semi-desert regions and on open plains. The Leopard carefully conceals itself in tall grass or other vegetation when stalking prey, or pounces on its victim from the branch of a tree, relying on its keen sight and excellent sense of hearing to detect other animals. In appearance very similar to the Jaguar, the Leopard is distinguished by its longer tail, its more graceful and slender body and by the absence of small, dark spots within the rosette markings. The Black Panther of Africa and Asia is really a form of Leopard.

CATS

The genus *Felis* contains most of the small Felidae which are properly termed cats – including the Wild Cat of Europe and the domestic cats. Like their larger relatives they have strong, padded paws with retractile claws, a round head and short muzzle. Apart from the ubiquitous domestic cat, the only member of this genus native to Europe is the Wild Cat, a forest-dweller found mainly in southern and central Europe, although it also occurs in Scotland and on the island of Corsica. The question of whether this cat is the ancestor of the domestic breeds is still under debate; it is probable that the common tabby cat at least is descended from the African Caffer Cat, a relative of the European Wild Cat and sometimes classed as a subspecies. The Caffer Cat was certainly domesticated by the Ancient Egyptians more than 4000 years ago, and spread to Europe from Greece and Rome, and into Asia via China.

Order Carnivora. Family Felidae. European Wild Cat (*Felis sylvestris*); male grows up to 65cm (26in) long, with a tail up to 35cm (14in) long. Found in Europe and parts of Asia, it inhabits upland woodland, forest edges and sometimes moors. It is yellowish grey with dark grey or black transverse stripes and in some places blotches; there are four longitudinal stripes on the forehead. Sometimes confused with feral domestic cats but its tail hair is thicker, and black at the blunted tip. Gestation lasts about 64 days and usually 4 young are produced once or sometimes twice a year. Caffer Cat (*F. lybica*) is found in Syria, and from Egypt south to the Cape.

CHEETAH

Unlike most of the other large Felidae which rely on stealth and surprise attack when hunting prey, the Cheetah depends on speed. It gives chase over open ground, having previously chosen its victim – usually a young or sick animal – from amongst the herds of grazing animals. The fastest of all land animals, the Cheetah can achieve a speed of up to 112 kilometres (70 miles) per hour, but it must capture its prey within a few seconds for it can maintain high speeds only for about 450 metres (500 yards). Its streamlined body is superbly adapted to this manner of hunting; the limbs are long, with semi-retractable claws which grip the ground, and its vision is extremely sharp. Cheetahs are sometimes confused with the Leopard, but the most obvious distinguishing feature however is the face markings: the Cheetah has two black stripes running from the corners of the eyes to the mouth.

Order Carnivora. Family Felidae. Cheetah (*Acinonyx jubatus*); males grow up to 210cm (84in) long, with a tail about 75cm (30in) long. Found in Africa and parts of Asia, although Asian specimens are extremely rare, they inhabit open plains, steppes and semideserts. Gestation lasts about 95 days with 2–4 young being born once a year. The female moves away from the troop as soon as the young are born. Young have a dark belly and flanks and long, grey mane-like hair on the head, back and tail. Diet consists mainly of hoofed mammals, especially Thomson's and Grant's Gazelles, although a variety of other animals is also taken, including hares, guinea fowl, warthogs and impalas.

LYNX

The lynxes are found in the northern hemisphere in both the Old and the New Worlds. They were once widely distributed throughout the mountainous forest regions of these areas; today they are extinct in many countries and rare in others, their numbers having declined owing to hunting (by man) and forest clearance. Like many of the Felidae they usually stalk their prey or ambush it from thick cover, making the kill by biting deep into the neck. These animals are distinguished by their short tail and by the long tufts on their ears. The Eurasian or Northern Lynx has a reddish-brown coat with whitish underparts and dark spots on the belly and limbs. The largest numbers occur in northern and central Europe and in Siberia and Mongolia, and a few are also found in France, Switzerland and Germany. A smaller race, the Pardel Lynx, occurs in Spain. North American representatives include the Canadian Lynx and the Bobcat.

Order Carnivora. Family Felidae. Eurasian Lynx (*Felis lynx*) grows up to 91 cm (36 in) long, with a tail up to 20 cm (8 in) long. It is found in northern and central Europe and parts of Asia and inhabits forested regions, usually in mountains. Diet consists of small mammals, birds and deer, and it will sometimes take sheep and goats. Canadian Lynx (*F. canadensis*), is found in pine forests in Canada and in the northern United States. Bobcat or Bay Lynx (*F. rufus*), is found in southern Canada, the United States and in parts of Mexico. Like other lynxes it hunts at night and feeds on rodents, hares, rabbits and small deer.

SPOTTED HYENA

Order Carnivora. Family Hyaenidae. Spotted Hyena (*Crocuta crocuta*); male grows up to 160 cm (64 in) long, with a tail up to 90 cm (36 in) long. Found in central and in southern Africa, it inhabits bushy and rocky country and savannahs. Gestation lasts about 95 days, with 1–2 and sometimes 3 young produced. Diet consists of fresh meat and sometimes carrion. The Striped Hyena (*Hyaena hyaena*) and the Brown Hyena (*H. brunnea*), are close relatives. The Striped Hyena is found in central Africa and parts of Asia, and the Brown Hyena only in South Africa. Both have erectile manes.

Hyenas have the worst reputation of all the carnivores, being accused of evil ways and cowardliness, and disliked for their offensive odour and unattractive appearance. In fact the hyena is an intelligent hunter that generally feeds on carrion only when live prey is scarce. It lives and hunts usually in packs, giving chase to large hoofed mammals and attacking them while they flee by biting their hind quarters and belly. The Spotted Hyena is the largest member of the family Hyaenidae. The coarse, rough coat is greyish to yellowish with dark brown spots; the tail is short and the front legs are longer than the back, producing a noticeable downward slope. In common with other hyenas it has a powerful jaw and extremely sharp teeth. These predators hunt at night, resting during the day in rock crevices or in their burrows.

RACCOON

Order Carnivora. Family Procyonidae. Common or North American Raccoon (*Procyon lotor*) grows to 70 cm (28 in long), with a tail up to 25 cm (10 in) long. Found in North America, it inhabits broadleaved and mixed woodlands. Despite being hunted for its fur and meat, the raccoon is still found in large numbers throughout the continent. Gestation lasts about 63 days, and 4 young are produced in a litter, of which there is only one each year. Diet consists of fruit, nuts, insects, birds' eggs, small birds, rodents, earthworms, frogs, crabs, mussels and fish. Crab-eating Raccoon (*P. cancrivorus*) is found in Central and South America.

Raccoons are New World carnivores. The Common Raccoon inhabits most of North America where it lives near water in both broadleaved and mixed woodlands. Easily recognized by its black facial mask and bushy, ringed tail, the Common Raccoon is not only terrestrial but climbs with agility and is an adequate swimmer. It is omnivorous and takes a wide variety of plant and aquatic animal food as well as birds' eggs, small birds and insects. Contrary to the old belief, the raccoon does not wash its food before eating but rather uses its hands to gather food from the water. A nocturnal animal, it emerges at dusk from its den which is usually made in a tree hollow or an abandoned burrow. The more northerly species hibernate during the winter although they may wake up and move about at intervals. A close relative found in South America, the Crab-eating Raccoon, has a slender tail, longer legs and shorter hair.

WOLF

Order Carnivora. Family Canidae. Wolf (*Canis lupus*) grows to 1.5 metres (60 in) long, with a tail up to 40 cm (16 in) long. It is found in North America, Asia and Europe – in Scandinavia, the Iberian Peninsula, parts of Italy and eastern Europe; it is probably now extinct in France. It inhabits forests, scrub or bushland. Mating takes place in the winter. The female lines the den with soft materials such as moss, leaves and hair from her belly. Gestation lasts about 63 days. One litter with 4–6 cubs (sometimes 8) is produced. The family remains together until the cubs are independent. Male helps to care for the young. Diet consists of large vertebrates, i.e. deer and caribou, and some plant food.

The Wolf belongs to the family Canidae which also includes jackals, foxes and both wild and domesticated dogs. It was once widely distributed throughout the northern hemisphere in both the Old and New Worlds. Owing mainly to persecution by man, the Wolf is now rare in Europe and its numbers are also declining in North America (where it is known as the Timber Wolf). In appearance and habits the Wolf exhibits many of the family characteristics: long, slim legs with non-retractable blunt nails, a long muzzle with sharp teeth and powerful jaws, and an exceptionally keen sense of smell. All of these are aids in hunting. The Wolf pursues its prey, giving chase until the victim drops from exhaustion. During the spring and summer, when both male and female are concerned with rearing their young, one or other of the parents will hunt alone; but in winter the family hunts as a group, sometimes in a small pack.

59

DOGS

The domestic dog's closest living relatives in the wild are the Wolf and the Jackal. Both of these have been considered at some time as the ancestor of the domestic breeds, for the domestic dog interbreeds with both, and in some cases is remarkably similar in appearance. However, most authorities now acknowledge that although the dog's genetic inheritance owes much to the Wolf, other members of the canine family have at different times, and in different places, contributed to modern breeds. Evidence suggests that dogs were used by man about 15,000 years ago in parts of the Mediterranean. Today some 400 breeds of dogs exist, many of which are very ancient. Of all domesticated animals, dogs are the most varied in size and appearance, ranging from the tiny Chihuahua to the gigantic Irish Wolfhound.

Order Carnivora. Family Canidae. Subfamily Caninae. Domestic Dog (sometimes classed as *Canis familiaris*). Size ranges from 15cm (6in) high (the Chihuahua) to 1 metre (40in) high at the shoulder (Irish Wolfhound). Bitches come into heat twice a year. Gestation lasts about 63 days, depending on the breed. There are usually 3–10 young, and sometimes as many as 20. Young become sexually mature at 10 months. Classes include the hunting dogs (Retrievers, Spaniels, Bloodhounds), Terriers, Pinschers, Schnauzers, Poodles (originally bred for coursing hares), German Shepherds, Spitzes and Huskies.

FOX

Widely distributed throughout the northern hemisphere, the adaptable, common Red Fox occurs in a variety of habitats from woodland to suburbia. It is also present in Australia, where it was introduced around 1850. Where ground cover is abundant the Fox makes its den in bushes, excavating a burrow or taking over the abandoned burrow of another animal; in human communities it will make its home under outbuildings and in cellars. Its diet is as varied as its habitat. The Fox will take advantage of whatever food is available, from small mammals and domestic poultry to fruit and berries. In spring the vixen gives birth to four or five young which are born blind and have a greyish woolly coat. The Fox has an ambiguous relationship with man. It is a pest to the farmer and a carrier of rabies, but in countries where fox hunting flourishes as a sport it is afforded some protection to ensure its survival.

Order Carnivora. Family Canidae. Red Fox (*Vulpes vulpes*; the North American race was once classed as a separate species, *V. fulva*) grows to 75 cm (30 in) long, with a tail 48 cm (19 in) long. Native to Europe, North Africa, Asia and North America, it was introduced into Australia. Vixen lines her den with soft materials. Gestation lasts about 53 days with one litter a year of 4–5 cubs. Diet consists of small mammals, including rabbits, voles, shrews and moles, and refuse, birds, berries, insects and fruit. Other fox species include the Arctic Fox (*Alopex lagopus*) of North America, Europe and Asia and the Fennec Fox (*Fennecus zerda*) which inhabits the deserts of North Africa and the Middle East.

BLACK BEAR

The largest of the carnivores belong to the family Ursidae, composed of seven species of bears which are found mainly in temperate and northern regions of the northern hemisphere. The Black Bear occurs only in North America, from Alaska south to the mountainous regions of central Mexico. In most respects it is a typical bear, exhibiting many of the family characteristics: a heavy body supported by short but strong limbs, feet with naked soles that end in five toes with sharp, non-retractile claws, excellent hearing and sense of smell but poor eyesight. Prior to winter it stores up food reserves in the form of body fat and then retires to a den for the whole of the winter. In common with other bears it does not undergo complete hibernation; its body temperature does not alter and it often wakens from sleep to feed. The female gives birth to the cubs towards the end of winter and remains with them in the den for two months.

Order Carnivora. Family Ursidae. Black Bear (*Ursus americanus*) grows up to 180 cm (6 ft) long. It is found in forests of North America from Alaska to central Mexico. The coat is usually black but sometimes brown. Females give birth every second year to 2 cubs which weigh about 453 g (1 lb) at birth. Diet consists of plant foods such as berries, conifer cones and roots, and animal food such as salmon, insects and small mammals; they are also extremely fond of honeycomb and will invade bees' nests. A close relative, the Brown Bear (*U. arctos*) is found in Scandinavia and the Pyrenees in Western Europe (although once found in most parts including Great Britain), eastern Europe, north and central Asia and North America.

POLAR BEAR

One of the most impressive of the bears, the Polar Bear is a solitary wanderer of the Arctic wastes, often swimming or travelling across the ice for many miles in search of food. Rarely found inland, it frequents the coastal regions and edges of the frozen seas. Its preferred food is seals, taken by stalking or by ambushing as they come up to an air hole in the ice to breathe. In summer the Polar Bear may feed on mosses, lichens and sea birds and their eggs. Unlike most other bears its sight is good and the soles of its feet are covered with hair, a feature that enables the bear to grip the smooth icy surface. Dens are made in the crevices of ice floes or in rocky outcroppings, but the pregnant female excavates a hole in the snow in which she bears her cubs. The mother remains in the den with her young for a considerable period; other Polar Bears are active the year round.

Order Carnivora. Family Ursidae. Polar Bear (*Thalarctos maritimus*) grows up to 2.7 metres (9 ft) long. Found in the Arctic regions of Europe, Asia and North America, it inhabits coastal regions and pack ice. The young remain with their mother for the first two years of life. Gestation lasts about 8 months, with usually two tiny young born in November or December. Coat is yellowish white, and the neck is longer than that of the Black and Brown bears. The most carnivorous of all the bears, its diet consists mainly of seal, also fish, birds and their eggs, and some plant food in the summer.

STOAT

Order Carnivora. Family Mustel-idae. Stoat (*Mustela erminea*) grows to an average 30cm (12in) long, with a tail up to 11cm (4½in) long. Males are markedly larger than the females. It is found in the northern parts of North America, most parts of Asia, in Europe (mainly in the Alps and Pyrenees and in central Europe) and in Greenland, as far north as the edges of the Arctic Ocean. Diet consists mainly of rabbits, also other small animals, especially ro-dents and birds. It will occasion-ally take insects and berries. It is active day and night.

Members of the family Mustelidae include stoats, weasels, mink, badgers and skunks, all characterized by thick, silky fur and the presence of anal glands which secrete a strong odour when the animal is alarmed. The Stoat is widely distributed throughout Europe, Asia and North America (where it is known as the Short-tailed Weasel or Ermine). In the most northerly parts of its range the Stoat grows a white coat in the autumn under its brown summer coat and, following the moult, has a pure white pelage with black-tipped tail: the 'ermine' that is so highly valued by the fur trade. Like the Weasel, to which it is very closely related, the Stoat is found in a variety of habitats including fields, woodlands, parks and human communities. Although the female bears only one litter a year, she will mate either in the spring or summer, with delayed implantation of the embryo resulting from the later mating.

MARTEN

Order Carnivora. Family Mustelidae. Pine Marten (*Martes martes*) grows to around 50cm (20in) long, with a tail up to 25cm (10in) long. Found in Europe (rare in the British Isles) and Asia, it inhabits broadleaved, mixed and coniferous woods. Diet consists of small mammals and birds, some insects, berries and fruit. Beech Marten (*M. foina*) grows up to 47cm (18.5in) long, with a tail up to 25cm (10in) long. Found in Europe (not in the British Isles) and Asia, it inhabits rocky terrain, woods, parks and gardens. Diet is mainly the same as the Pine Marten's. American Marten (*M. americana*) inhabits forested regions of North America. Diet similar to that of the above two species. The Sable (*M. zibellina*) is found in Asia.

Martens occur in most parts of Europe, Asia and North America. Many are forest-dwellers that hunt their prey in trees. Like other members of the Mustelidae they have long, slender bodies and short legs and feed primarily on flesh. Well-known species include the Pine Marten and the Beech or Stone Marten (both found in Eurasia) and the American Marten of North America. The last is sometimes call the American Sable but it is not the same species as the Sable of Asia which is hunted for its very valuable fur. Both the American and Pine Martens inhabit woodland, where they climb trees to prey on squirrels and birds. The Beech Marten is more closely associated with man and is often found in or near human communities. Martens have an exceptionally long gestation period for such small mammals, normally over 240 days, owing to the delayed implantation of the fertilized egg.

EURASIAN BADGER

Order Carnivora. Family Mustelidae. Eurasian Badger (*Meles meles*) grows up to 72.5cm (29in) long, with a tail up to 18.7cm (7¼in) long. Found in most of Europe, with the exception of the far north, and in Asia as far east as Japan. It inhabits open fields, moors, woodland, scrub, rubbish tips, hedgerows, and sometimes human communities. Colour of pelage may vary from reddish-brown to black on the upper parts and from white to yellowish on the underparts. Its senses of smell and hearing are good but its eyesight is poor. Diet is omnivorous and includes small mammals, earthworms, snails, insects, fruits, nuts and grass. American Badger (*Taxidea taxus*).

Apart from the well-known and common Eurasian Badger, distinguished by its black and white striped head, six other species are found in Europe and Asia, and one in North America. The Eurasian Badger is nocturnal and crepuscular (active at twilight). It excavates elaborate underground tunnels and chambers with several entrances, usually made in loose, sandy soil in areas with ample ground cover. It will also occasionally make its home under buildings. A fastidious creature, the Badger periodically replaces its bedding of plant material (with which it lines its sleeping chamber) or brings it up into the sun to air it. Scrapes made away from the burrow, or set, are used to deposit faeces. A male (boar) and female (sow) usually occupy a set, together with any young. As with the Marten, implantation of the fertilized egg is delayed and gestation occupies some seven or eight months.

OTTER

Once quite common throughout its range, the Common or Eurasian Otter is becoming increasingly rare and in some areas is extinct. The main contributing factors to its decline are the pollution of waters and persecution by man. Both a terrestrial and aquatic animal, the Otter is a superb swimmer and diver and on land can cover short distances as fast as a man. Underwater the Otter propels itself with its long, tapering tail while on the surface it paddles with its forelimbs. Movement on the ground is with a bounding gait or, especially in snow, by sliding. Otters are predominantly nocturnal and maintain territories near still and flowing waters, concealing themselves during the day in hollows, burrows or amongst vegetation such as reed beds. Adults are generally solitary; groups consist of a mother and her young which stay with her for the first year.

Order Carnivora. Family Mustelidae. There are around 17 species of otter distributed throughout Europe, Asia, Africa and North and South America. Eurasian Otter (*Lutra lutra*) grows up to 82.5cm (33in) long, with a tail up to 55cm (22in) long. Found in Europe and Asia and in North Africa, it inhabits lakes, rivers, marshes and streams. One litter of 2–4 young is born at any time of the year after a gestation period lasting about 63 days. Diet consists of fish, crustaceans and aquatic insects. Burrows are made in banks, with the entrance below the water. North American Otter (*L. canadensis*).

DOLPHIN

Order Cetacea. Suborder Odontoceti. Superfamily Delphi-
noidae. Family Delphinidae,
true dolphins. Common Dolphin
(*Delphinus delphis*) grows up to
2.4 metres (8 ft) long. It is found
in most parts of the world but
especially in the warm waters of
the Pacific and Atlantic Oceans,
the Indian Ocean, the Mediterranean and Black Seas, where it
gathers in schools. Diet consists
of fish, cuttlefish and squid.
Common Porpoise (*Phocaena
phocaena*) grows up to 1.7
metres (5½ ft) long. Found along
the coasts of the northern hemisphere, it lives in small schools.
Diet consists of fish, crustaceans
and squid.

The superfamily Delphinoidea is the largest group of toothed whales and includes among others the killer whales, dolphins and porpoises. True dolphins (Family Delphinidae) are often confused with porpoises but are distinguished by their long, narrow beak, more slender body and by the shape of the dorsal fin which is sharply-pointed, whereas in the porpoise it is blunt and more triangular. The Common Dolphin has a dark brown back with white underparts, and pale and dark bands along the flanks. In common with most other species it is an excellent swimmer, sometimes reaching speeds of up to 48 kilometres (30 miles) per hour, and when spouting will leap clear out of the water. Dolphins are thought to be the most intelligent of all animals, with a highly developed system of communication and social structure. Their attention to, and care of their young, the sick and the wounded are well-known.

WHALES

The large group of marine mammals of the order Cetacea, which are generally termed the whales, are the most marine of all the mammals yet they are probably descended from terrestrial animals that returned to the sea millions of years ago. Whales have a torpedo-shaped body, well-developed forelimbs and a large, powerful tail ending in two horizontal flukes. The hind limbs are absent, as are external ears, and the nostrils have been replaced with a blow-hole situated at the top of the head. The Cetacea are divided into two large groups: the Odontoceti or toothed whales and the Mysticeti or baleen whales. Baleen whales are distinguished mainly by the huge plates of baleen or whalebone that hang from the upper jaw. These are used to filter food from the water. The toothed whales comprise the largest and most varied group, with some 80 species ranging in size from the huge Sperm Whale to the small porpoises.

Order Cetacea. Suborder Mysticeti – composed of 12 species of baleen whales. Blue Whale (*Balaenoptera musculus*) is the largest animal ever known to exist. This whale can grow to a maximum length of 30 m (100 ft) and a weight of 150 tonnes. It is found in all the waters of the world but mainly in the Antarctic Ocean. Once one of the most widely-hunted whales, today it is a protected species. Gestation lasts about 12 months. Single calf, born every second year, weighs about 2.5 tonnes at birth and measures about 7 metres (23 ft) in length. Like all baleen whales it feeds exclusively on Krill. Suborder Odontoceti includes Killer Whales, dolphins, and porpoises (see page 69). Sperm whale (*Physeter catodon*) grows up to 20 metres (66 ft) in length.

WILD HORSE

It is probable that the only wild horses alive today are those kept in captivity, for there has been no evidence of their existence in the wild for some years. ('Wild horses' such as the Mustangs of North America and the horses of the Camargue are feral horses.) Once widely distributed over the plains of Europe and Asia, the last species to be found in the wild was the Mongolian or Przewalski's Horse, discovered in the remote mountains along the Mongolian/Chinese border in the late 1800s by the explorer Colonel Przewalski. From early cave paintings it is known that wild horses have been associated with man since the beginning of his history; at first they were hunted for meat but were later tamed and harnessed to carry goods and people. Indeed this association was largely responsible for the ultimate extinction of the wild species. Hunting and competition for grazing land contributed greatly to its decline.

Order Perissodactyla — odd-toed ungulates. Family Equidae, composed of one genus *Equus* to which the horse, ass and zebra belong. All are monodactyl or single-toed. Wild Horse (*Equus caballus przewalskii*), 1.35 m (4.5 ft) high to the withers, has a reddish-brown coat, white muzzle and short ears. The coat becomes long and shaggy in winter. The mane is black, short and erect and the tail long and black. It has reddish callosities or 'chestnuts' on the inside of each leg. This horse was formerly found in the Takin Shara-nura mountains of Asia. The European Wild Horse, the Tarpan (*E. caballus gmelini*), was found in Europe east to southern Russia. The last Tarpan to be found in the wild died in 1879. Today there are re-bred specimens in several of the world's zoos.

HORSES

Order Perissodactyla. Family Equidae. Domestic Horse (*Equus caballus*). Modern horses retain many of the characteristics of their wild ancestors: long limbs and a single toe protected by a horny nail or hoof which has a cushion or pad to soften impact with the ground; these are adaptations for running at speed. All Perissodactyls have a simple digestive system which is compensated for by strong molars which are used to grind down the hard tissues of grasses and thereby make them more digestible. The long neck enables horses to graze without lowering their body and the sharp front teeth are designed for clipping off tough stalks.

Some 50 million years ago the ancestor of the modern horse lived in tropical forests in Europe and North America: *Hyracotherium* was a small, dog-sized animal with three toes on the hind feet and four on the front. Its descendants, living primarily on the plains of North America, evolved gradually into animals very similar to those known today, the more advanced forms making their way into Asia. The first people to domesticate and make use of the horse were probably nomadic tribes of the Asian steppes who later introduced these horses into the Middle East. The first attempts at breeding horses were aimed mainly at producing stronger and heavier animals for use in warfare but since then a great variety of breeds have been developed for many purposes. Some of the best-known modern breeds include the enormous Shire of England, the Scottish Shetland Pony and the Quarter Horse of the United States.

ASIATIC WILD ASS

Order Perissodactyla. Family Equidae. Asiatic Wild Ass (*Equus hemionus*); its height to shoulder varies from 100 to 140 cm (40–55 in). Found in Central Asia, it inhabits arid and semi-arid regions. The Kiang (sometimes classed as *Equus kiang*) is the most numerous. It lives in herds of 5 to 300 individuals, usually comprising females and their young led by an old mare. The stallions generally keep apart from the herd. Mating begins in August when the stallions fight for possession of the females. Gestation lasts for 335 days, with one young born in July or August. Diet consists of grasses and other plants.

The wild asses of Asia, once numerous over the arid regions of Central Asia, are, like most of the Equidae, diminishing in number. Today they are found in the mountains of Tibet and in the low-lying deserts of Mongolia, India and the Middle East. Smaller than horses, they also have larger ears, a short, erect mane and tufted tail. All are extremely swift runners. Superficially they differ mainly in size and the colour of their coat. The largest is the Tibetan Wild Ass, or Kiang, which lives in herds on the mountain plateau. Its summer coat is reddish with a dark stripe down the back, but in winter the coat becomes browner, thicker and longer. Those dwelling in the lower regions include the Mongolian Wild Ass or Kulan, the Onager of Syria and Iran, and the Ghor-khar of India. Some authorities classify all these equids as subspecies of *Equus hemionus*, others classify the Kiang as a separate species.

AFRICAN WILD ASS

The sole surviving representative of the African Wild Ass, the ancestor of the domestic donkey, is the Somali Wild Ass, today found only in the northern part of that country. It is highly likely that this animal will shortly face the same fate as the Nubian Wild Ass which has become extinct in the last few years, for only a few Somali Asses still exist in the wild. Larger than its Asian relative, the African Wild Ass is greyish in colour, has longer ears and narrower hooves which enable it to travel in stony regions. It is generally agreed that neither the Somali Wild Ass nor the Nubian Wild Ass is the ancestor of the domestic donkey. Opinion now favours the theory that the donkey is descended from some extinct African species or race. Domesticated donkeys bred for specific purposes include the Savoy Ass, raised for climbing in Alpine regions, and the Macedonian Ass of the Balkan countries.

Order Perissodactyla. Super-family Equoidea. Family Equidae. African Wild Ass (*Equus asinus*). The only surviving race, the Somali Wild Ass, is found in Somalia. Feral asses are found in northern Africa, and southern Europe. The domestic donkey is used primarily for carrying goods and is able to travel over difficult terrain where other animals could not easily pass. These animals will breed with horses. Crosses between the two include the Mule, a cross between a male ass and a mare, and the Hinny which is a cross between a stallion and a female donkey.

ZEBRA

Three species of zebra are found in Africa, the most numerous and widely distributed being the Common Zebra. Like Grevy's Zebra and the Mountain Zebra, its numbers are rapidly dwindling. The first two species dwell in the open plains where they live with other grazing ungulates such as gnus, migrating hundreds of kilometres with them in the dry season to areas where there is a permanent supply of water, and returning in the rainy season to bear their young. Various races of Common Zebra are found throughout its range, each with different patterns of stripes, the tendency being for the more northerly races to have broader stripes. Common Zebras are sometimes erroneously called Burchell's Zebra; this is an extinct species from which the Common Zebra probably descends. The Mountain Zebra is found only in the mountains of South Africa. It is the only equid with a dewlap on its throat.

Order Perissodactyla. Suborder Hippomorpha. Family Equidae. Common Zebra (*Equus burchelli*) grows from 120 to 140 cm (4–4½ ft) high at the shoulders. It is found south of the Sahara from the north-east to the south-west of Africa. Grevy's Zebra (*E. grevyi*) grows to 160 cm (5.3 ft) at shoulder height. Found in Kenya and Somalia, this is the largest zebra and the most densely and narrowly striped; the belly is unmarked. Mountain Zebra (*E. zebra*) has a shoulder height of 118 to 132 cm (3¾–4⅓ ft). It was the first species to be scientifically named and described. Two races are distinguished, both of which are rare: Cape and Hartmann's Zebra. Distinguished by their long ears, narrow hooves and heavy head. Like Grevy's Zebra, the Mountain Zebra has an unmarked belly.

WHITE RHINOCEROS

Order Perissodactyla. Suborder Ceratomorpha. Family Rhinocerotidae. White Rhinoceros (*Ceratotherium simum*), measures 440 cm (175 in) long and is up to 195 cm (78 in) tall. Weight 4000 kg (8800 lb). Two races are usually distinguished: *C. simum simum*, confined to South African nature reserves, and *C. simum cottoni*, found in part of Uganda, the Congo and the southern Sudan. Gestation lasts about 18 months. The single young is not weaned for a year and remains with its mother for several years. Diet consists of grasses. Black Rhinoceros (*Diceros bicornis*) grows up to 370 cm (147 in) long from head to rump, males may be 170 cm (67 in) tall at the shoulder. Weight up to 2000 kg (4400 lb).

Five species of rhinoceros, three in Asia and two in Africa, comprise the family Rhinocerotidae. The largest member, and the second largest land animal after the elephant, is the White Rhino which inhabits Central Africa and Natal in South Africa. In reality it is a greyish colour (the word white comes from the Afrikaans *weit*, meaning wide, which is used to describe the broad mouth). The White Rhino has the thick skin, massive body and horns characteristic of the family. The horns are a product of the dermis; they are not true horns but a mass of agglutinated hairs. Unlike its close relative, the Black Rhinoceros, which is predominantly nocturnal, solitary and a browser, the White Rhino is a grazer and often lives in small groups. Because the horns are valued as an aphrodisiac, particularly in the Far East, their hides used for leather goods and their flesh eaten, both have been hunted and are now in danger of extinction.

WILD BOAR

The Wild Boar is found in most parts of Europe, in Asia and in North Africa, and inhabits woodland with dense thickets. Males are generally solitary but the females and the young, led by an old sow, form herds known as sounders. The ancestor of the domestic pig, the Wild Boar has an elongated head ending in a snout, thick skin covered with bristles and a long, straight tail. In the male the upper canines are long and grow out of the mouth, curving upwards towards the top of the head to form tusks. The coat of the Wild Boar is dark brown in the adult and grows longer during the winter; the young have a yellowish-brown coat with long stripes from the shoulders to the rump. Like the domestic pig, it is omnivorous and roots in the ground for much of its food, of which it takes an enormously wide variety. Most spend the day in shallow scrapes under the cover of bushes and roam about during the night.

Order Artiodactyla – even-toed ungulates. Suborder Suiformes. Family Suidae, Old World Pigs. Wild Boar (*Sus scrofa*) grows up to 1.8 metres (6 ft) long, and the tail to 20 cm (8 in) long. Found in Europe (except Britain, Ireland, Iceland and the north) and in Asia and North Africa, it inhabits woodlands. They mate from November to January, and gestation lasts up to 140 days. The female makes a scrape in the ground and lines it with leaves. There is usually one litter a year of 4–12 young. Diet includes small animals and birds' eggs, and plant food such as mushrooms, roots, nuts and cultivated crops.

PIGS

Order Artiodactyla. Suborder Suiformes. Family Suidae. Domestic Pig (*Sus scrofa domesticus*) has been developed in Europe and Asia from the Wild Boar. Gestation lasts 115 days and the sow bears two or three litters per year, each with 6–20 young. British breeds include the Black Berkshire and the White Yorkshire; the latter was originally bred from a Siamese hog and a Yorkshire breed in the 18th Century. Raised for its meat, it is a rapid grower; after one year it weighs about 118 kg (260 lb) and when fully-grown 305 kg (672 lb). Two Asian forms are the Masked Pig of northern China and the Vietnamese Pig. England and the United States are among the main pig-farming countries of the world.

Domestic pigs are descended from the Eurasian Wild Boar. They have much in common with their wild ancestors, including a natural inclination for cleanliness. Their habit of wallowing in mud serves the purpose of ridding their skin of external parasites and cooling the body. Like the wild species, they are omnivorous and will eat anything, rooting in the ground with their long muzzles. For this reason they are prone to a number of internal parasites which, in the past, gave rise to religious prohibitions in certain countries against eating pork and pork products. Modern methods of pig-farming have eliminated these infestations. Pig breeds have been developed mainly for their meat and one of the prime considerations in breeding is to produce an animal that will grow rapidly, although fertility and good milk production are also important factors in sows that are raised as breeders.

BUSH PIG

Bush pigs are found in tropical forests in most parts of Africa south of the Sahara to the Cape. Like other wild pigs they gather in droves or sounders, but in this case a large boar usually dominates the group on which he imposes a strict discipline. The boars are massive and ferocious both in appearance and fact, and courageously defend the drove against predators on the slightest provocation, although in most circumstances they will give warning by grunting before they attack. Although the tusks are shorter than those of the Wild Boar (see page 78), they are nevertheless dangerous weapons, capable of inflicting fatal wounds on dogs and similar-sized animals, and even occasionally on the Bush Pig's greatest natural enemy, the Leopard.

Bush Pigs vary enormously in the colour of their coats but the majority are dark brown to black. The coat is covered in coarse bristles which grow densely on the back, neck and shoulders to form an erectile crest or mane of white and black hairs. The face has white or whitish markings and boars have obvious protuberances on the top of their muzzle below the eyes. The ears are sharply pointed and have long tufts of hair at the tips. Although they are mainly nocturnal, in the remoter parts of Africa these pigs will often move about during the day. Ordinarily, however, they take cover among tall grasses and reeds and emerge only at dusk. During their nightly forages Bush Pigs travel in a group and may cover considerable distances to find a suitable feeding ground. When they become alarmed, the drove stampedes through the forest, crests erect and uttering noisy grunts. Sows usually bear their young during the rainy season, and prior to the birth construct a nest of grasses among the grass or bush. The newly-born young are strikingly marked with yellow and brown stripes. Although Bush Pigs cause considerable damage to crops by rooting in the earth, the same activitity aerates the soil and removes or destroys grubs and larvae which are harmful, thereby contributing to the welfare of the habitat.

Order Artiodactyla. Suborder Suiformes. Family Suidae. Bush Pig (*Patamochoerus porcus*) grows about 75cm (30in) tall with tusks up to 15cm (6in) long. Found in Africa, south of the Sahara, it inhabits tropical forests, usually near water. Diet is omnivorous and includes roots, berries, fruit, crops, insects, eggs and snakes. Sow bears 5–6 young, sometimes as many as 10. Young are born mainly during the wet season but births may take place at any time of the year. They live in sounders of 5 to 20 individuals.

HIPPOPOTAMUS

By the rivers, lakes and swamps of Africa lives the amphibious Hippopotamus. It bathes in the water during the day and emerges at night to travel away from the banks and shores to feed. Superbly adapted for its way of life, the Hippo's legs and feet aid it in swimming, and walking on the river bottom or over muddy swamps. Its ears, nostrils and eyes are so placed on its broad head that while the rest of its body is completely underwater these sense organs can still function to listen for, scent and sight danger. By closing its nostrils and ears and taking huge lungfuls of air, it can totally submerge for up to ten minutes. The Hippo's aquatic habits not only protect it from would-be predators but also from the heat of the sun, to which its hairless skin is particularly sensitive as over-exposure causes it to become dry and crack.

Order Artiodactyla. Family Hippopotamidae. Common Hippopotamus (*Hippopotamus amphibius*) grows up to 430 cm (172 in) long with a tail 50 cm (20 in) long. Weight 2–3 tonnes, sometimes up to 4. It inhabits rivers, lakes and swamps and is found in Africa south of the Sahara. Gestation lasts about 235 days after which a single young is born, usually on land but occasionally in the water. Vegetarian diet includes grass and leaves. The skin is greyish, sometimes with pinkish undertones. The hippo has a symbiotic relationship with certain birds and fish which rid its hide of parasites. It contributes to the growth of aquatic plants through its excretions which add valuable nutrients to the water.

CAMELS

Two species of camel are native to the desert regions of the Old World; the two-humped Bactrian Camel of Asia and the single-humped Arabian Camel of south-west Asia and North Africa. Long associated with man, and often termed the 'ship of the desert', the Arabian Camel was domesticated around 1800 B.C. and the Bactrian Camel some thousand years later. These animals have been of great economic importance to nomadic tribes for, as well as being beasts of burden, their flesh and milk provide food, their hair and skin provide clothing and their droppings are used as fuel. Although camels have a high tolerance of dehydration, contrary to the old belief, they do not store water in their humps (these act as food reserves). In the dry hot season camels have to drink every two or three days, although in the cold Asian winters the Bactrian Camel can go without water for several weeks. A few of these camels still exist in the wild.

Order Artiodactyla. Suborder Tylopoda. Family Camelidae, composed of camels and llamas (see page 84). Bactrian Camel (*Camelus bactrianus*) grows 2.10 metres (7ft) high to the top of the humps. Found on the cold deserts of central Asia, and a few still exist in the wild in the Tibetan deserts. Heavier and stockier than the Arabian Camel its coat is thick and long in winter, and is shed in summer. Arabian Camel (*C. dromedarius*) has a height to the hump of 2.25 metres (7½ft). Native to south-west Asia and North Africa, it was introduced into Australia, southern Africa and southern Asia. Its build is more slender, and its legs longer and thinner than in the Bactrian Camel. The Dromedary is a race of Arabian Camel developed for riding and racing.

LLAMAS

Order Artiodactyla. Suborder Tylopoda. Family Camelidae. Llama (*Lama glama*) has a height to its shoulders of 1.2 metres (48 in). It is found in highlands of Peru, Chile, Ecuador and Argentina, and on the Altiplano of Bolivia. Only the castrated males are used as pack animals; the others are pastured and raised for breeding. Alpaca (*L. pacos*) has a height to the shoulders of up to 90 cm (36 in). It inhabits the Altiplano along the borders of Bolivia and Peru. Vicuna (*L. vicugna*) inhabits remote mountainous regions of Peru. Guanaco (*L. guanicoe*) inhabits the mountains and plains of western South America.

Four species of llama are found in South America: the Vicuna, Guanaco, Llama and Alpaca. Only the first two are now found in the wild; the ancestors of the domesticated Llama and Alpaca died out many centuries ago. Like their larger, humped relatives, the true camels, the Llama and the Alpaca have played an important role in the lives of local peoples. Although its fur, flesh and skin are utilized by Peruvian Indians, the Llama is predominantly a pack animal, capable of carrying a load of up to 45 kilograms (100 pounds) for several days over rough mountainous terrain. Their woolly coats, intermixed with stiff guard hairs, provide a coarse fabric suitable for rugs and blankets. The smaller, sturdier, long-haired Alpaca however is kept for its wool which is fine and light. Of the two wild species the Guanaco is the most widespread. The smaller Vicuna is more timid and is restricted to Peru.

PERE DAVID'S DEER

The existence of this deer was first brought to the attention of the western world in 1865 when the French missionary and naturalist, Father Armand David, visited the Emperor of China's garden near Peking. The descendants of a few specimens brought to Europe are the only surviving members of the species, for in 1890 floods destroyed most of the herd in China and the few that escaped were killed a few years later. Nothing is known of their original habitat, but it is probable that they were inhabitants of wet, marshy areas, for they feed mainly on aquatic vegetation and are good swimmers. Their broad hooves suggest an adaptation for living on marshy ground. Unlike other Cervidae, Père David's Deer have a long tail and the back antlers are straight and slender while those at the front are forked. The coat is reddish-grey with white underparts and there is a white ring around the eyes.

Order Artiodactyla. Suborder Ruminantia – even-toed ungulates with ruminating or cud-chewing habits and usually a four-chambered stomach. This group includes sheep, goats, cattle and giraffes. Père David's Deer (*Elaphurus davidianus*) grows to about 120 cm (48 in) at the shoulders. Originally from Asia, it is now found in zoological gardens and parks in Europe, especially Woburn Abbey in England, from where specimens have been sent to Austria and South America and released into the wild. They mate in June or July, and gestation lasts about 10 months. A single young is produced. Diet consists mainly of aquatic plants. Summer antlers are dropped in November. Winter antlers are developed by January and quickly dropped.

FALLOW DEER

Originally from the eastern Mediterranean and south-west Asia, Fallow Deer now occur throughout Europe and have been introduced into Australia, North and South America and parts of Africa. Today they are found predominantly in parks and zoological gardens, the largest number of those found in the wild consisting of feral forms. Colouration of the coat may vary from white to dark brown but most commonly it is bright fawn with obvious white spots in summer, fading in winter to grey with faint markings. The tail is black above and white beneath and there are black and white markings on the rump just above the tail. The Fallow Deer can be distinguished most readily from other species by the antlers which, when fully developed, have large palmate areas. In remote places these deer are diurnal, but otherwise they rest during the day and begin to feed at dusk.

Order Artiodactyla. Suborder Ruminantia. Family Cervidae. Fallow Deer (*Dama dama*) has a height to the shoulder of up to 95cm (38in). Native to eastern Europe and south-west Asia, it is now found throughout Europe, in Australia and North and South America, and inhabits woodland and parks. Rutting season is from October to November. Gestation lasts about 230 days. Single young (rarely twins) is born in early summer. Diet includes grasses, young shoots of broadleaved trees, nuts and berries. Antlers are present only in the males. In all the Cervidae the antlers are bony outgrowths and not horns. They are usually branched and are shed each year.

RED DEER

Order Artiodactyla. Suborder Ruminantia. Family Cervidae. Red Deer (*Cervus elaphus*) males grow up to 137cm (54in) high at the shoulders with antlers up to 120cm (4ft) in length. Native to Asia, North Africa and Europe, and introduced into New Zealand and North America, it inhabits forest margins, moors, plains and scrubland, from lowlands to above the tree-line. Rutting season is from September to October. Like most deer, the stags fight for possession of the females. Gestation lasts about 238 days. Single young is born in May or June. Diet includes grasses, small shrubs, herbs and mushrooms. A close North American relative, the Elk or Wapiti (*C. canadensis*) is found mainly in Canada but also in some parts of the United States.

The Red Deer belongs to the genus *Cervus* which contains the largest number of cervids. It inhabits open grassland, moors and scrubland in semi-arid regions but prefers the marginal areas between forest and plain. The Red Deer occurs from the most westerly regions of Europe east to China, north as far as the Arctic Circle and south to North Africa. It has also been introduced into New Zealand and North America. The coat is reddish in summer, also varying from dark brown to yellowish. In winter it is commonly brown and the hairs are long and dense. The rump is white from the tail to the top of the legs. Adult males have large antlers which may have from 8 to 16 tines on each. When fully developed the antlers are hard but while they are in the process of growth they are soft and covered with a hairy skin called 'velvet'. Mature adults usually shed their antlers in early spring and by July possess a full-sized new pair.

EURASIAN ELK

Order Artiodactyla. Suborder Ruminantia. Family Cervidae. Elk (*Alces alces*) has a height to shoulders of 2.5 metres (8½ft). Found in northern Europe, Asia and North America, it inhabits marshy, swampy areas near broadleaved woods. Females and young keep together in small herds during the summer months. They are joined by males in the rutting season in September and remain together during the winter. Males fight for possession of the females and often become dangerous to humans as they are known to charge anything that moves. Gestation lasts 225 days. One to two young (sometimes three) are born in May. Diet consists of leaves and aquatic plants.

The largest of the Cervidae, the Eurasian Elk or American Moose once inhabited much of Asia and the forests of central and western Europe. Although it died out in most parts of Europe and was restricted largely to the Scandinavian countries, in recent years it has begun to increase in numbers and distribution and is now found in parts of Germany, Austria, Poland and Czechoslovakia. In North America the Moose occurs in its largest numbers in Canada and Alaska, and in parts of the western and eastern United States. (This species should not be confused with the North American Elk, which is related to the Red Deer; see page 87.) The Elk has a greyish-brown coat, long legs and a very short neck which inhibits its ability to graze. It is therefore mainly a browser on land, but it also feeds on aquatic plants found on or just below the surface of water. In the summer it is found near (and in!) rivers and lakes.

REINDEER

Order Artiodactyla. Suborder Ruminantia. Reindeer (*Rangifer tarandus*) has a shoulder height of up to 120 cm (48 in). Found in Arctic regions of Europe, Asia and North America, it inhabits the tundra, taiga and mountainous regions. Herds are generally small and consist of females, their young and the young males. Rutting takes place in September and October during the southerly migrations. Gestation lasts 224 days with one or two calves being born in May as the herds move north for the summer. The diet includes mosses, lichens, bark, leaves and young shoots.

Found in the northern regions of both the Old and New Worlds, Reindeer, or Caribou, are the only deer that migrate over long distances and the only deer to be domesticated. Although some authorities distinguish two species – the European Reindeer and the North American Caribou – most classify all members as one species. In North America the Caribou move north in the spring and spend summer on the tundra, returning south to the forests before winter sets in. In Eurasia Reindeer have been domesticated for hundreds of years and few are found today in the wild. Northern peoples, especially the Lapps, have in the past relied on them for most of their food, clothing and transport. Unlike other deer, both sexes possess antlers although the male's are stronger than the female's ; as in other species they are shed annually. Reindeer are very similar to Red Deer in appearance but their coat is greyish-white to brown.

ROE DEER

One of the smallest of the cervids, the Roe Deer occurs in both coniferous and broadleaved woods, in forest margins and fields. Almost tailless, when alarmed it erects the hairs on its rump so that it appears to possess a ball of fluff on its hind quarters, and flees with long jumps into the scrub or bush. The colour of its coat varies from sandy to reddish in summer, with a white rump patch in the females and a pale yellow patch in the males. The winter coat is greyish to dark brown or black with pale underparts. During the rutting season, from July to August, the bucks mark out part of their territory as a rutting area, fraying the bark of trees with their antlers and distributing a secretion from the scent glands on their foreheads on to the vegetation. The does do not bear their young until the following May, as the fertilized egg is not implanted into the wall of the womb until December or January.

Order Artiodactyla. Subfamily Ruminantia. Family Cervidae. Roe Deer (*Capreolus capreolus*) males grow up to 75 cm (30 in) high at the shoulders, with antlers up to 30 cm (12 in) long. Found in Europe and Asia but absent from Ireland, Wales, parts of the Mediterranean region and northern Scandinavia, they inhabit coniferous and broadleaved woods, forest margins, fields and meadows with ample cover. Their presence can sometimes be detected by 'roe rings' in the earth made around trees and shrubs. Rarely found in large groups, most Roe Deer live in small family groups in winter. In summer does and fawns stay by themselves, and bucks are solitary but remain in the vicinity. Both browsers and grazers, diet is mainly of leaves, grasses and herbs.

OKAPI

This fascinating and unusual animal is found only in the tropical forests of the Congo. Although it has been hunted by the Pygmy people for thousands of years, it has been known to science only since 1900. The Okapi, although as large as a horse and with stripes resembling the Zebra's, is nevertheless related to neither and is in fact a close relative of the Giraffe. Like the Giraffe the Okapi has an extremely long, prehensile tongue which it uses to clean its face and ears and, with its lips, to gather leaves and twigs from high branches. Timid and retiring, the Okapi is seldom seen in the wild; it generally lives alone among dense vegetation where its protective colouring keeps it well camouflaged. Characteristically of the ruminants the male has two 'horns' which are short, bony and, except at the tips, covered with skin and incline towards the back of the head. Its hearing is extremely acute but sight and smell are not well developed.

Order Artiodactyla. Suborder Ruminantia. Family Giraffidae. Okapi (*Okapia johnstoni*) has a height to the shoulder of 150–160 cm (60–64 in), with a tongue 35 cm (14 in) long. It is found in the tropical rain forests of the Congo between the Ituri and Uele Rivers. Male and female remain together for 2–3 weeks in the breeding season. Gestation is long, lasting up to 449 days. Single young is produced. Diet consists of twigs, leaves, fruit, and also insects. As in the Giraffe, the front limbs are slightly longer than the hind limbs, and the skin is thick. The name 'okapi' comes from the Pygmies' name for this animal. It is also named after its discoverer, Sir Harry Johnstone.

GIRAFFE

The Giraffe is the tallest of the mammals. It is free from competition with other animals for food, for its enormously long neck allows it to browse on leaves and twigs far out of reach of other terrestrial creatures. The Giraffe is an inhabitant of the tropical savannah where acacia trees are plentiful. Its long tongue and lips enable it to delicately gather foliage from among the tree's thorns. The Giraffe drinks, and occasionally grazes, by straddling its legs so that it can lower its head to the water. Giraffes are often seen in fairly large groups of up to 30 individuals, most of which are females and their young. They live peacefully among herds of Zebra, Gnus and other grassland dwellers such as the Ostrich. The Giraffe has no enemies apart from man and the lion, and an adult is perfectly capable of defending itself from the latter by kicking out savagely with its powerful legs. However, in the face of danger it usually flees in haste.

Order Artiodactyla. Family Giraffidae. Giraffe (*Giraffa camelopardalis*) males grow up to 5.4 metres (18 ft) tall, females to 4.5 metres (15 ft) tall, with a tail over 90 cm (36 in) long, and a tongue 42.5 cm (17 in) long. It is found on the savannahs of tropical Africa. Usually 12 races are distinguished mainly by the patterns and colours of the coat. Gestation lasts about 450 days, producing a single young about 180 cm (72 in) tall at birth. Diet consists of leaves, twigs and sometimes grass. In spite of the length of its neck, the Giraffe has only 7 neck vertebrae, like most other mammals. Both male and female have two short horns covered with skin. Vocalizations include whistles and gurgles.

AFRICAN BUFFALO

Some 46 genera comprise the family Bovidae (ruminants with hollow horns which are never shed, and two toes on each foot). The family includes sheep, goats and antelopes as well as the Old World or true buffalo. Only one species of buffalo is found in Africa, although variations occur in size and shape of the horn. The most common race is the Cape or Black Buffalo of South and East Africa; and the smallest, the Dwarf Forest Buffalo or Bush Cow of Central and West Africa. Buffaloes inhabit areas where there is an abundance of grass for grazing, and water and forest cover near the grazing grounds where they can rest during the day. Although Buffaloes have few enemies and are capable of defending themselves with their sharp horns and powerful legs and feet, they generally take to the water when threatened. Those animals that are captured by the Lion are the cows and calves who have strayed away from the herd.

Order Artiodactyla. Suborder Ruminantia. Family Bovidae. African Buffalo (*Syncerus caffer*) grows up to 1.5 metres (5ft) high at the shoulders. Dwarf Forest Buffalo (*S. caffer nanus*) grows up to 107cm (43in) high at the shoulders. Found in Africa south of the Sahara, they inhabit forest edges near savannah; most are now on reserves Diet consists of grasses. Both male and female have massive horns; the coat is generally black, and the hair very short. They have nocturnal habits, drink at least twice a day, and take mud baths to rid themselves of skin parasites.

DOMESTIC CATTLE

Order Artiodactyla. Suborder Ruminantia. Family Bovidae. Aurochs or Wild Ox (*Bos primigenius*) was once found in Europe, Asia and North Africa. Banteng (*B. banteng*) is found in South-east Asia, Java and Borneo. Zebu (*B. indicus*) is found in India. Domestic cattle, like most ruminants, have a four-chambered stomach and a complex digestive system that enable them to digest the tough fibres of grasses on which they feed. The stomach consists of the rumen and reticulum from which partially digested food is regurgitated; ruminated food then passes into the third stomach, the omasum, and from there into the true stomach or abomasum.

Most authorities now agree that the ancestor of domestic cattle was the Wild Ox or Aurochs of Europe, Asia and North Africa. No longer found in the wild, this bovine became extinct when the last survivors died out in Poland during the 17th century. Today its descendants are found on most man-made grasslands and cultivated natural grasslands throughout the world. The many breeds of domestic cattle include those raised for milk production, such as the Jersey, and those bred for beef such as the Aberdeen Angus and Highland Cattle. Wild cattle in Africa and Asia have also been used by man as draught animals or providers of meat and milk. The Banteng of South-east Asia has been domesticated in Java and Bali and is sometimes interbred with the Zebu, a humped cow with a large dewlap under its neck. In India the Zebu is the sacred cow of the Hindu religion.

NORTH AMERICAN BISON

Order Artiodactyla. Suborder Ruminantia. Family Bovidae. North American Bison (*Bison bison*) males grow up to 180 cm (72 in) high at the shoulders. Originally found on grasslands throughout Canada and the United States, today it is a protected animal that is restricted to national parks and reserves. Colour of coat is generally brownish with darker, longer hair on the head, neck and feet. It can be distinguished from other species by the large hump on the back. Gestation lasts about 230 days, producing a single young, usually every one or two years. Three hours after birth the calf is able to run with its mother. Diet consists mainly of grass and sometimes bark.

Known in North America as the Buffalo, this bovine is related to the European Bison which is now extinct in the wild, and during the 19th century it almost suffered the same fate as its Old World relative. The American Bison once roamed the extensive grasslands of North America and although it was hunted by the Indians for food, clothing and housing materials, it was not killed in sufficient numbers to endanger its survival. When Europeans began to explore and settle on the continent, however, they began to slaughter the Bison in enormous numbers and by the late 19th century of the 75 million animals present before their arrival, only about 500 remained. The American Bison is a massive creature with long, thick dark hair and short, pointed horns present in both males and females. Two races are distinguished, the Plains Buffalo on the prairies and the Wood Buffalo at higher altitudes.

GNU

Order Artiodactyla. Family Bovidae. Brindled Gnu (*Connochaetes taurinus*) has a height to the shoulders of up to 145cm (58in) and is found in southern Africa. White-bearded Gnu (*C. taurinus albojubatus*) is found in East Africa, especially on the Serengeti Plains National Park in Tanzania. Black Wildebeest (*C. gnou*) has a shoulder height of 120cm (48in), and is found only in protected areas of South-West Africa. Horns are present in both male and female gnus, but those of the male are heavier. Females usually give birth to a single young at the beginning of the wet season.

These curious-looking antelopes, with their elongated face, cow-like horns, long, plumed tail and shaggy mane, are called wildebeest (wild cattle) by the Afrikaans. Two species are found in Africa: the Black Wildebeest or White-tailed Gnu and the Blue Wildebeest or Brindled Gnu, of which one other race, the White-bearded Gnu, is distinguished. The most numerous and widespread, the Brindled Gnu occurs in southern and East Africa where it inhabits the savannahs, usually associating with large herds of zebra, gazelles and other grazing animals. Twice a year they migrate in their hundreds – in search of water during the dry season and back to the open grassy plains in the wet season when the rains have stimulated the growth of plants. At this time the females usually give birth. Today the Black Wildebeest is found only in private parks and on game reserves in South-West Africa.

SPRINGBOK

The Springbok or Springbuck is a beautiful antelope of South Africa and is that country's national emblem. Until the end of the 19th century, millions of these animals grazed on the open plains and periodically undertook mass migration during which many were trampled to death by the constantly moving herd or drowned in the rivers. The reason for these movements is unknown but may have had something to do with reducing numbers in over-populated areas. In smaller numbers the Springbok also makes seasonal migrations from the highlands where it spends the summer, to lower regions where it feeds during the winter. The Springbok is distinguished from all other antelopes by a fold of skin which runs from halfway down its back to its tail. When the antelope becomes agitated or excited, the fold opens to reveal a long, white crest of hair which stands erect, and the Springbok leaps into the air.

Order Artiodactyla. Family Bovidae. Springbok (*Antidorcas marsupialis*) has a shoulder height of 68–90 cm (27–36 in). Horns up to 40 cm (16 in) long, are lyre-shaped and are present in both male and female, although the female's are shorter and narrower and not as curved. The coat is fawn above and white below with a long brown stripe dividing the upper parts and the white belly. The face has two black stripes running from the ears to the muzzle. Gestation lasts about 170 days, usually producing a single young in October or November, although it may be born at any time of the year. Diet consists of grasses and leaves.

IMPALA

One of the best-known and the most graceful of the antelopes, the Impala is a wonderful jumper, leaping about 3 metres (10 feet) into the air over a distance of 9 metres (30 feet). Both a grazer and a browser, it occurs on acacia savannahs and along forest margins and in dense thorn bush; it lives in fairly large herds of between 10 and 100 individuals. The horns, which are found only on the male, are long and curve upwards and outwards in a lyre-shape, coming slightly in at the tips. The coat is fawnish-red on the upper parts and white below, with black vertical stripes on either side of the rump which is white below the tail. Black tufts of hair are present on the back of the hind legs above the hooves. In the rutting season the rams often fight viciously and, when courting the ewe, will erect and fan out their tail over their back.

Order Artiodactyla. Family Bovidae. Impala (*Aepyceros melampus*) can be up to 85–98cm (34–39in) high at the shoulders. The horns grow to about 67cm (27in) long. Found in South and East Africa, it inhabits open grassland and dense thorn bush. Gestation lasts 180 days, producing a single young, but sometimes twins, born between November and December. Diet consists of grasses, leaves and twigs. It needs to drink regularly and is therefore always found near permanent sources of water. Predators include the Leopard, Cheetah and Wild Hunting Dog.

THOMSON'S GAZELLE

With its fawn and white coat, black stripe along its flanks and lyre-shaped horns, Thomson's Gazelle is remarkably similar to the larger Springbok (page 97) of South Africa. It also has the habit of 'pronking' or jumping several feet into the air when agitated but it lacks the erectile crest along the back and does not arch its body when it leaves the ground. Like the Zebra, with which it often associates, Thomson's Gazelle travels hundreds of miles from well-watered regions, where it spends the dry season, to the lush pastures of the open plains, sometimes travelling 15 kilometres (9 miles) to reach the areas where the first of the rains are falling. During the wet season the pregnant females leave the herd prior to giving birth and remain with their young for about three months. The fawns are darker than the adults and attempt to hide from predators by lying flat and motionless on the ground.

Order Artiodactyla. Subfamily Ruminantia. Family Bovidae. Thomson's Gazelle (*Gazella thomsoni*) males grow to about 65cm (26in) high at the shoulders, with horns up to 32cm (13in) in length. Found on the open plains of East Africa, it associates with the Zebra (page 76), the Impala (page 98) and Grant's Gazelle (*Gazella granti*). During the rutting season the mature males gather harems; usually the young males form bachelor herds, and older males are solitary. Diet is of grasses of which they choose the tender new shoots. Animals that graze together tend to eat different parts of a plant so that there is an adequate supply of food for each species and the land is not overgrazed.

CHAMOIS

The Chamois belongs to a relatively small group of bovines that are intermediate between the antelopes and goats and in fact are sometimes referred to as goat-antelopes. It is an extremely agile climber and jumper that inhabits mountainous regions between 760 and 2285 metres (2,500 and 7,500 feet) above sea level. Its horns, present in both sexes, are unlike those of any other ruminant; they are straight along most of their length but curve backwards into a hook at the tips. During the summer the Chamois feed high in the mountains, often above the snow-line, but in winter when food is scarce and temperatures drop they move down the mountains to forested regions, often as far as the valleys. At this time they undergo a moult, their reddish summer coat being replaced with a black coat with white patches on the flanks, head and neck. The rut is in November, with the young being born in the spring.

Order Artiodactyla. Suborder Ruminantia. Family Bovidae. Chamois (*Rupicapra rupicapra*) has a shoulder height of 80–85 cm (32–34 in) It is found in south-west Europe, in the Alps, Apennines and the Carpathian Mountains and in the highlands of south-east Europe and south-west Asia. Nine races are usually distinguished. Gestation lasts 180 days, producing a single young, and rarely twins. Diet includes alpine plants; in winter also mosses, lichens, bark and coniferous leaves. In North America, the Rocky Mountain Goat (*Oreamnos spp*), occupies a similar habitat and is similar in appearance. It is found in the Rocky Mountains from Alaska to Montana.

MUSK OX

Order Artiodactyla. Suborder Ruminantia. Family Bovidae. Musk Ox (*Ovibos moschatus*) has a shoulder height of up to 1.5 metres (60 in). These bovines get their name from the musky odour emitted from their bodies. The coat is dark brown with buff patches on the shoulders and lower legs, and may be up to 90 cm (36 in) long in places. The feet are splayed, enabling the Musk Ox to travel over ice and snow without sinking. Females calf every two years and bear one young. Diet consists of stunted willow, moss, lichen and bark. Found in Arctic regions of Canada and Greenland, they are now protected animals.

The Musk Ox is found on the freezing Arctic wastes of Canada and Greenland. It survives only in small numbers, having been almost exterminated during the last century by explorers and fur traders. In a land that affords no natural shelter, Musk Oxen have only their long, thick coats and each other to protect them from the cold; they huddle together and share the warmth given off by their bodies. This mutual defence against the climate is also extended against predators, for when they are threatened they form a circle with their heads and sharp horns pointing outwards so that they can attack the enemy head on. The young are placed in the centre of the phalanx. Musk Oxen live in herds of up to 100 individuals and apart from man their only natural enemy is the Wolf. The Eskimos, who have hunted them for hundreds of years for meat and clothing, have probably not contributed to their decline.

IBEX

Order Artiodactyla. Suborder Ruminantia. Family Bovidae. Alpine Ibex (*Capra ibex ibex*) males grow up to 85cm (34in) high at the shoulder, with horns up to 75cm (30in) long. Females are considerably smaller, with horns which are small and point straight up, less than 20cm (8in) long. Gestation lasts about 147 days, and females give birth once a year to 1–2 young. Diet consists of grass and alpine plants such as lichens and mosses. Spanish Ibex (*Capra ibex pyrenaica*) is found in the mountains of Spain but is now extinct from the Pyrenees. It is greyish brown with a black dorsal stripe along the back and pale underparts.

The Ibex belongs to the genus *Capra*, a group of bovines that are often given the common and general name wild goats. Two European races of *Capra ibex* are the Alpine Ibex native to the Alps and the Tatra Mountains, and the Spanish Ibex, once found in the Pyrenees and other mountainous regions of Spain, but now extinct in the former area. Today both forms are rare in the wild but attempts to re-establish the Alpine Ibex in parts of Switzerland have been successful. The males are impressive creatures with long, curved horns that have thick, transverse ridges along their length; the coat is greyish to dark brown and there is a small beard under the chin. Does, their young, and the young bucks form herds, the older males only joining them in the breeding season. Throughout the year the Ibex lives above the tree-line, and like the Chamois (page 100) is an excellent climber and jumper.

DOMESTIC SHEEP

Like the goat, the sheep was one of the first animals to be domesticated and used to provide meat and clothing for ancient civilizations. Evidence suggests that the history of its association with man goes back to 6000 B.C. The modern wool industry began in the 14th century in Spain, for the Spaniards were the first to breed sheep on a large scale for their wool. One of the better-known breeds, still raised today for its fine, high-quality wool, is the Spanish Merino. Adapted to living in dry climates, it was introduced into Australia in the 1800s. It is now found in many of the other great sheep-farming countries of the world and has been cross-bred with other stock to produce fine breeds. Domestic sheep are thought to be descended from the Argali, the largest of the wild sheep, which is found in the arid regions of central Asia.

Order Artiodactyla. Suborder Ruminantia. Family Bovidae. Argali (*Ovis ammon*). The Domestic Sheep is classed as *Ovis aries*. The Merino is distinguished by its horns, found only in the rams, which curl into a spiral and are situated at the side of the head. Like other Woolly Sheep it has a thick, fluffy coat that lacks the bristles found in the group known as Hairy Sheep. In Britain breeds of sheep are classified as long-wools and shortwools; the former have white faces and lack horns. One example is the English Leicester which has been crossed with the Merino to produce excellent wool breeds. The Southdown, an English breed common in the United States, Canada and other Commonwealth countries, has a black face.

DOMESTIC GOAT

Order Artiodactyla. Suborder Ruminantia. Family Bovidae. The domestic goat is a descendant of *Capra hircus aegagrus*, and is more prolific than any of the wild forms, giving birth to 2–3 kids. Its milk yield is at least twice that of the cow. The Swiss Saanen, a white and hornless goat, is raised in England, the United States and many other countries. Other popular breeds are the Toggenburg, Nubian and Anglo-Nubian. The Kashmir is native to Tibet and has a fine, silky undercoat that produces an equally fine wool called cashmere. The Angora comes from Turkey and in this case it is the outer coat that is used in the manufacture of wool.

Domesticated by the 7th century B.C., goats were raised to provide meat, milk and milk products, leather and, in some cases, wool for ancient tribes. Although they are not as intensively farmed today in the western world as sheep and cattle, domestic goats give a higher yield of milk than cows in relation to their size, and stock such as the Kashmir and Angora provide some of the finest and softest wool of any animal. However most goats are raised for their milk which provides valuable nutrients for people who are unable to drink cow's milk, for it is more digestible and does not carry tubercular infections. Among the modern, favoured breeds are the Swiss or Alpine goats, the Wallis, the White German and the Toggenburg.

MOUFFLON

Order Artiodactyla. Suborder Ruminantia. Family Bovidae. Moufflon (*Ovis musimon*) can be 65–75cm (26–30 in) high at the shoulders. Native to Corsica and Sardinia, it is now found in France, Germany and Czechoslovakia. Gestation lasts 150 days, and usually a single young is born in early spring. It generally lives in small herds led by an old ewe. Males fight violently during the breeding season. Similar species found in North America is the Bighorn Sheep (*O. canadensis*), found from Alaska to Mexico in wild and remote mountainous regions. It also occurs in north-eastern Siberia. The so-called wild sheep of Scotland (found only on the islands of St Kilda) are feral sheep.

The Moufflon occurred originally only on the islands of Corsica and Sardinia, where it was almost exterminated by hunters before steps were taken to protect it. In recent years it has been introduced into several European countries including France, where it has most successfully adapted to a new habitat. The Moufflon is the only living species of European Wild Sheep. Its natural habitat is rocky mountain forests but elsewhere it lives in high alpine regions and in broadleaved and mixed woodlands. Only the rams possess the wide spiralling horns once prized by trophy hunters; in the female the horns are lacking or, when present, are shorter and only slightly curved. In winter the ruddy coat becomes darker and the pale patch or 'saddle' on the back becomes a more brilliant white. The stocks introduced into continental Europe are generally larger and heavier than those native to the islands, where food is less abundant.

AFRICAN ELEPHANT

Only two species of elephant, the African and the Indian (or Asiatic) make up the order Proboscidea which once contained over 352 distinct species. The African Elephant, the larger of the two and the largest of all living land animals, is divided into two races, the Bush Elephant and the Forest Elephant, the former being the most widely distributed and the latter being confined to tropical forests of west Africa. The African Elephant is distinguished by its flat forehead, concave back, large ears and ringed trunk; both the males and females have long tusks. It lives in family groups led by an old female (the matriarch) who dominates the females, calves and young bulls. These groups join up with others and the group of adult bulls and form a large herd of some 30 to 50 individuals. Elephants are most active during the cool hours of early morning and evening. They rest in the shade at noon and sleep, sometimes standing, at night.

Order Proboscidea. Suborder Elephantoidea. Family Elephantidae. The African Elephant (*Loxodonta africana*) grows to approximately $11\frac{1}{2}$ metres (38ft) at shoulder height and the adult bull can weigh up to 6 tonnes. The tusks can be up to 3 metres (10ft) long and weigh about 55kg (121lb) each. Once found throughout Africa south of the Sahara, they are now found mostly in game parks and reserves. Gestation lasts about 21 months; the single young is not weaned until it is about 5 years of age, and may live up to 60 years. Diet consists of grass, fruit, bark, twigs and roots. It can cause considerable damage to vegetation by uprooting trees. It drinks about 130 litres (228 pints) of water each day. Once hunted extensively for its ivory tusks, it is now a protected species.

ASIATIC ELEPHANT

Order Proboscidea. Suborder Elephantoidea. Family Elephantidae. Asiatic Elephant (*Elephas maximus*) grows up to 3 metres (10ft) high to the shoulder, and weighs about 4 tonnes. Tusks grow up to 1.8 metres (72in) long, each weighing between 4.5–9kg (10–20lb). It is found in southern Asia and Sumatra. Several races are distinguished, including those from Sumatra, Malaya, Sri Lanka and Bengal. It lives for up to 70 years. Gestation lasts 21 months; during birth the pregnant female is attended by other females who act as 'midwives'. Elephants are renowned for the care and attention they give to each other. They help their injured, and gather around their dying, trying to raise them to their feet.

In the teak forests of Burma and elsewhere in Asia the Asiatic or Indian Elephant has been used as a working animal for thousands of years. It is the species most frequently seen in zoos and performing in circuses. In India, where it is used in religious ceremonies, it is considered a sacred animal. Smaller and darker than its African relative, the Asiatic Elephant also has smaller ears, a smooth trunk, a domed forehead and an arched back. The tusks are shorter and, in the cows, they are not externally obvious or are totally absent. In the Sumatran race both sexes lack tusks. The elephant's trunk has evolved from the upper lip and the nose and is used as a hand, an arm, and as a siphon for holding water. These massive creatures bathe frequently to rid themselves of external parasites and insects. For the same reason they often take mud baths.

BIRDS

Birds fascinate man for two reasons. Firstly, there is a great variety of them differing in appearance and behaviour; and secondly, birds are masters of the air – the one element that man has always desired to conquer. The fact that these animals are aerial creatures is not enough, however, to justify their classification as a separate class, Aves. Many insects possess wings and can fly, butterflies and moths being prime examples, and of course, among the mammals, bats have taken to the air.

The general characteristics common to all Avian members, including the ostrich and penguin which cannot fly, is that they are warm-blooded (the only modern group to be so besides the mammals), two-legged vertebrates with forelimbs ('arms') that have evolved into wings. They also have thin skin that is covered with feathers or, on the legs and feet, with scales. The last feature, scales, is the most obvious of those inherited from the birds' ancient ancestors, the reptiles. Scales, the beak and feathers are all composed of a material called keratin. All three – beak, scales and feathers – form in, and are products of, the epidermis and dermis (the skin).

All the characteristics mentioned above are intricately related and appeared during the slow process of evolution as a response to the need to survive in as efficient a way as possible. For example, bird feathers evolved from reptilian scales – one way in which birds adapted to their environment.

Evolution of Birds

Between 225 and 180 million years ago the cold-blooded animals that dominated the earth were faced with the problem of keeping warm. As the external temperature rose and fell, so did the body temperature of the animals. During cold periods they became sluggish and inactive, and were therefore less able to defend themselves against predators. Some animals solved the problem by developing feathers, fine structures which trap air and insulate the body. These enabled the animals to maintain a constant temperature, a temperature produced by the assimilation of food in the body which also gave it the energy to remain active. They became, in other words, warm-blooded animals.

The process by which these primitive animals evolved into present-day birds is, of course, complex and occurred over millions of years. Exactly how they evolved is still a mystery. Some indication of what the first bird-like creatures were like was revealed by *Archaeopteryx* (meaning ancient wing), a winged, feathered animal regarded by many as the first bird. Its fossil remains were discovered in 1861 at Solnhofen in Germany, a discovery that established for the first time the link between ancient reptiles and birds. *Archaeopteryx* was not able to fly although it possessed feathered forelimbs. The skeleton of solid bones was far too heavy to enable it to take off from the ground or remain in the air for long. It used its clawed toes and three clawed fingers to climb up trees; from there it glided from tree to tree. The fossil also shows that *Archaeopteryx* had a long, bony tail and had teeth in the upper and lower jaws.

Archaeopteryx lived about 150 million years ago during the Jurassic Period. By the beginning of the Tertiary Period (from 64 to 2 million years ago), many of the modern birds – including kingfishers, herons, storks, pigeons and gulls – had evolved. Almost all had lost the teeth, bony tail and clawed fingers although they had, and still retain, vestigial digits at the tip of the wings. It took another 62 million years for all the present-day families and most of the 8000 known species to evolve. During the same period many of the primitive species became extinct, and by the Eocene Epoch, some 50 million years ago, the birds' body had become an aerodynamic masterpiece with the lightness and the strength required to lift it off the ground and power it through the air.

Flight

Apart from the ratites, or running birds (see

page 115), all birds have a wide breastbone (the sternum) which extends at the top into a keel-shaped structure known as the carina. Strong pectoral muscles which control and power the wings are attached to the carina. The penguins, although they are flightless, also possess the carina and breast muscles which, in their case, power the wings which paddle the body through water. Unlike *Archaeopteryx*, modern birds have hollow porous bones, many of which have become fused making the body compact and light. Lightness is further gained by the lack of teeth which over millions of years have been replaced by a horny beak containing the nostrils, and a large internal cavity behind the small lungs where the air sacs are found. The real function of the air sacs is unknown although they definitely aid the bird in flight. Unlike most mammals, birds reproduce by laying eggs. This characteristic – oviparity – is inherited from reptiles, and increases the birds' flight efficiency as carrying an embryo inside the body would considerably restrict movement through the air.

Most of the bird's internal organs are placed by the wings near the body's centre of gravity. This helps to maintain balance and stability. The wings are responsible for keeping the bird airborne. The upper surface of the wing is convex (curves outward) so that air passes over it more quickly than it does over the inward-curving (concave) lower surface. The difference in the rate of air movement makes the pressure greater underneath the wing than it is on top, thus creating a vacuum above which sucks the bird up and gives it lift.

Wings, however, are useless to birds unless they are feathered. Feathers streamline the body and, most important, on the wings and tail give the necessary surface to support flight. Feathers are produced in the skin, in small buds from which the shaft grows. The shaft is hollow at the base and is fed by blood vessels until the feather is fully grown. Thereafter the feather can be called a dead structure, in the same way that fingernails are dead once they reach the tip of the finger. By means of minuscule hooks, the branches of a feather lock together, making it a cohesive unit and so producing a smooth surface.

Flight feathers on the wings (remiges) and the tail (rectrices) are known as the quill or vane feathers. The remiges are responsible for getting the bird through the air; they are found on the 'hand' and forearm, point backwards and are covered at the base with smaller feathers known as wing coverts. The length of the wing quills may vary from one group of birds to another and, together with the length of the forearm, which is also variable, give the wing its shape. In eagles, for example, the outer wing quill is longer than the others, which are shorter nearer the body. This, together with a long forearm, gives the eagle its characteristic slow but powerful wing beat.

The rectrices, which constitute the 'tail', are also covered at the base by coverts and are used for steering and braking. The feathers which cover the body and produce its streamlined shape are the contour feathers or tectrices. In most chicks they are preceded by down, the very fine soft feathers that are also present in adult birds under the contour feathers. Other fine, hair-like feathers are called filoplumes; these are present on most parts of the body.

Once in the air, most birds can fly in two ways: by flapping or beating their wings and by gliding, the earliest form of flight. The manner in which a species flies depends very much on its way of life, to which the actual shape of the wing is adapted. Woodland birds have rounded wings perfect for short, fluttering flights through foliage; ocean-going birds, such as the albatross, have long, wide wings on which they can glide for hours, assisted by air currents (see page 118). The hummingbirds have the most intricate flight patterns; they can fly forwards, backwards or hover in the air by moving their wings in complex loops and arcs. All this movement and activity however requires a considerable amount of energy which necessitates the constant intake of a great deal of high-energy food and an extremely efficient digestive system that can deal with it as quickly as possible.

Feeding

Birds are voracious eaters and take a wide variety of food: seeds, grains, fruit, flower heads, nectar and fungi; insects, their larvae and eggs; crustaceans, fish, molluscs; and small mammals and rodents. Some also eat other birds and their eggs. The Herring Gull even eats the young of its own species. Most species specialize in particular kinds of food although some, such as the House Sparrow, have extremely wide tastes. Obviously the more varied the diet, the more likely the bird is to survive during food shortages. Many perching birds eat insects and other small animals in the spring and summer and feed on berries in the autumn and winter when the former are scarce but the latter are abundant. The Wren, however, restricts its food to insects and spiders, and many of these small birds die of starvation during harsh winters. Yet, whatever the type of food, almost all of it contains large amounts of protein or carbohydrates which the body converts into heat and energy.

The intake of food begins of course with the beak, a horny, keratinous structure of fused plates, which extends from the upper and lower jaws to form the upper and lower mandibles. To some extent the beak has taken over the function of teeth for, although it is not used to masticate, it is sharp enough to grasp, tear meat and crush the shells of seeds and crustaceans. The particular shape of the bill is indicative of the type of food a bird eats: the Goldfinch is a seed-eater and has a narrow, pointed bill with which it extracts seeds from flower heads and catkins; the Shoveler feeds in shallow waters and has a long, flattened beak with fine plates known as lamellae along the edges which are used to strain tiny animals from mud and water. The considerable use to which the beak is put causes it to be constantly worn away, and to compensate for this the tip of the beak grows continually.

Most birds use their tongue, which is highly mobile, only to manipulate food from the beak to the oesophagus, but the woodpecker's tongue is specially adapted for gathering small insects and their larvae from bark crevices.

The tip of the long, pointed tongue has small hooks and the whole surface is covered with a sticky substance to which the insects adhere.

Many birds are able to store food temporarily in a section of the oesophagus but generally the food passes through the body as rapidly as possible. In some birds the oesophagus contains an enlarged pouch, or crop, which softens food before it enters the stomach, or, in the case of male Emperor Penguins and pigeons, produces a milky curd-like substance on which newly-hatched chicks are fed for the first few days of life. Only a few species, such as juvenile Puffins and Emperor Penguins, store food in the form of fat.

From the oesophagus the food passes into the stomach, which is in two parts: the glandular stomach which secretes digestive juices and the muscular stomach, or gizzard, which mechanically grinds up food, often aided by small pebbles and grit which some birds swallow. The food matter then passes into the intestines where it is absorbed into the body and the waste is excreted through the rectum. Some birds regurgitate indigestible matter through the mouth in the form of pellets.

Senses

Little is known about a bird's sense of taste (although we know that some species can distinguish between sweet and sour, saltiness and acidity) and even less is known about their sense of smell. Some vultures can smell carrion from a distance and probably use this ability as well as sight to locate dead animals, but this is not necessarily true of all members of the group. In contrast it is known that a bird's senses of hearing and sight are extremely acute and are of primary importance not only in hunting but also in the detection of predators.

Vision may be of two types: monocular, in which each eye forms a separate image so that the bird can look in two different directions at once, or binocular, in which the two images formed by each eye overlap to form a single image which is three-dimensional. This gives the bird a means of measuring distance. A

bird's way of life affects its visual requirements. In hunters or birds of prey, for example, sight is predominantly binocular so
that they can spot their prey from a considerable distance and capture it with astounding accuracy. Monocular vision is a
characteristic of many small wood-dwelling
species, enabling them to find food easily and
at the same time keep an eye out for predators.

With the exception of owls, birds' eyes are
set towards or at the sides of the head, giving
them a visual range of about 300 degrees.
Owls, whose eyes are placed in the front of the
head, are able to turn their heads round as
much as 280 degrees to compensate for the
narrow range of their vision. In common with
some other species they also have sense organs
in the form of bristle feathers around the beak
with which they can identify objects at close
range.

Apart from nocturnal species, all birds are
able to distinguish colours in fine detail and
many male birds take advantage of this by
displaying their brightly coloured plumage in
threat or courtship display. This use of colour
is a warning to competitors that are infringing
on their territory or an enticement to the
female to mate.

Plumage

Territorial and sexual display are two functions of colourful plumage; another is identification, both of other members of the same
species and of the opposite sex. Incorporated
in the definition of a species is the fact that all
members of the group have the ability to
interbreed and reproduce without threatening
the continuation of their common genetic
inheritance. To do this they must first be
capable of recognizing those with whom it is
genetically 'safe' to breed. The colours and
patterns of the plumage provide strong visual
signals which birds use to differentiate between members of the same group and members of another group. The fact that colouration also plays an important part in distinguishing one sex from another is obvious in
the Mallard and most other Anatidae, where
the drakes are so brightly marked and the

ducks drab and relatively inconspicuous. In
some species, however, the sexes appear to be
very alike, but there is usually some characteristic, however small, that differentiates
male from female. Cock and hen Great Tits
share the same greenish upper parts, black
head and yellow underparts and both have a
black band running down the centre of the
breast, yet the band is broader in the male
than it is in the female. It is worth noting that
the male's courtship behaviour includes display of his breast to the female.

The actual colour of a feather arises from
one of two factors: from its chemical or from
its physical makeup. Some feathers are coloured because they contain certain pigments,
such as melanin, that produce reds, browns
and blacks; others are actually colourless but
appear coloured because the structure of the
feather is such that it interacts with light. Many
of the iridescent colours, such as those in a
Starling's plumage, are produced in this way.
Pigmentary or chemically-coloured feathers
are constant in their colour value, while
structurally or physically-coloured feathers
change according to the angle and amount of
light which they receive. The plumage colouration of many birds is produced by both factors.

It is essential that the plumage is kept in
prime condition so that flight and other
functions can be carried out as efficiently as
possible. Feathers must be kept clean, and
those that have worn through use, or which
have been lost or damaged must be replaced.
Most birds preen their feathers, using the
beak to smooth them and to distribute oil
along them from the preen (uropygial) gland
situated in the skin near the tail. This oil
ensures that the plumage is waterproof, an
essential condition for waterfowl and other
birds that live in watery habitats. A few birds,
such as the Cormorant, have feathers that
become thoroughly drenched so that the air
trapped in the feathers escapes. The bird loses
its buoyancy and becomes heavier, allowing it
to dive to a great depth. Some birds, including
the Heron, have special down feathers on their
breast which fray at the tips into a fine
powder; the birds distribute the powder down

through the plumage as a cleansing agent. Others take dust baths which serve the same purpose.

Worn feathers are replaced during the moult which may occur once or twice a year depending on the species. In most cases the feathers are lost and replaced gradually so that flight is not impaired, but some birds such as the Mallard undergo a complete moult at the end of the breeding season when they are virtually grounded for a few weeks until the new feathers appear. A second or partial moult takes place before nesting begins in the spring when a smaller proportion of feathers is renewed. Often the species which undergo two moults are the ones which have a brightly coloured plumage during the breeding season but, following the complete moult, become duller and less conspicuous.

Breeding

The breeding season is a time of great activity when the rituals of courtship and the rearing of young put additional strain on the bird, increasing the demands for energy, food and the need to protect itself and its young from predators. It is not by accident that reproduction takes place when food is abundant enough to supply both the parents and their ravenous young. For most species in the temperate regions of the northern and southern hemispheres breeding occurs annually in spring and summer. The remainder of the year is a time of sexual inactivity when they are free to wander in search of food. In tropical zones where the climate and availability of food remain more or less constant, birds often breed several times at regular intervals throughout the year, interspersed with periods of rest. In a few species the most advantageous times for nesting coincide with periods of scarcity for other animals. For example, the Marabou Stork of Africa raises its young during the dry season when drought brings starvation and death to many animals, thereby providing carrion for both the bird and its young.

Given normal conditions then, chicks are not likely to die from starvation; the danger to a brood comes from other sources. Because the fertilized egg develops outside the mother's body, it is particularly vulnerable to climatic changes and predation. Birds are able to compensate, however, for the loss of an entire clutch in several ways: a species may lay and raise more than one clutch in a season, and most which have a single brood can lay a replacement clutch if necessary. Nests are made under thick cover high above the ground or are camouflaged in some way. Hornbills, for example, take drastic steps to ensure that their eggs and young are safe from predators. The female incubates and raises the young in a tree hole which has been, except for a tiny opening, blocked up completely by the male and female before the clutch is laid (see page 179).

For birds which make annual journeys from winter quarters to their breeding sites, mating is usually preceded by the males selecting and marking out a territory which they defend against intruders. The territory may have several purposes: it may be used solely for display or it may also be used for feeding, mating and nesting. Birds which are resident in their area throughout the year generally remain in the same territory during the breeding season.

Song, display flights and exhibition of the plumage are used by the males to attract a female and to indicate that they are ready to mate, but display is not only a male prerogative. Among the Great Crested Grebes, for example, both male and female perform an elaborate ritual involving the mutual exchange of nesting materials, head-shaking and loud cries.

Probably the majority of birds are monogamous for at least one breeding season. The ostriches and some other types of birds, however, are polygamous, the male breeding with several females, and in very few instances the female taking more than one mate. Storks and swans are two well-known birds that mate for life, and both male and female share in nest-building and rearing the young. The degree of participation by the male in rearing the young varies considerably from species to species; some only mate, then leave the female to carry out the other tasks. The male Emperor Penguin, however, incubates the single egg continuously for 64 days, holding it on his feet under a fatty fold of skin on the belly. After the chick has hatched he feeds it for several days on a milky substance produced in his crop. The female then appears for the first time since laying the egg and only then does the male return to the sea to feed.

For most birds the process of mating is a precarious event as most males lack an external reproductive organ. A few, however, including ratites and ducks, have a primitive form of penis which becomes erect prior to copulation. In all other males the semen is emitted through the cloaca, the single external opening at the base of the body through which faeces are also passed. The act of copulation takes place when the male's and female's cloacae are placed together, this lasts only a few seconds and may be repeated several times.

Once the birds have mated, the fertilized egg spends approximately two days in the female genital tract, passing down from the ovary to receive the white (albumen), shell membrane and shell before it is laid. Some birds begin the incubation immediately after the first egg appears but others may wait until the clutch is complete. The incubation period varies from 11 days in some species up to a maximum of 81 days (in some of the albatrosses), the males of some species taking their turn at brooding. When the chick is ready to hatch it picks at the shell with its egg-tooth, a small pointed structure on the upper mandible, which disappears shortly after hatching. Most chicks are fledged by the autumn and can then hunt and fend for themselves, although a few species with very long juvenile periods, such as the Wandering Albatross, do not fledge for almost a year.

Seasonal Movements
At the end of the breeding season most birds move away from their nesting site. During these seasonal movements birds may cover vast distances over several continents or they may merely move to a different habitat within the same region. The Arctic Tern, which travels farther than any other bird, flies some 17,700 kilometres (11,000 miles) from the Arctic to the Antarctic at the beginning of the northern winter, returning again in the spring. The European Robin is not a migrant but it usually moves in the autumn from its woodland nesting site to parks and gardens.

The question of how birds actually find their destination is complex and much still remains a mystery. Most ornithologists are agreed that different species use different methods. Birds that travel over land are probably able to identify major landmarks and may also navigate by the sun, if they travel by day, or the stars if they travel by night. Yet there are cases of some birds unfailingly finding their way under abnormal circumstances which cannot be explained. It is possible that birds possess navigational skills that are far more sophisticated than any of those used by man.

Species and Orders
The Class Aves contains about 30 orders of birds. (The number may be several more or less, depending on whether some of the species, about which there is a difference of opinion concerning classification, are placed in separate orders or included in existing ones.) The 93 species included here are representatives of 19 of the best-known orders. The largest proportion are perching birds (Passeriformes). This order contains more species than any of the others. All of the species are living birds although, unfortunately, some of them are in danger of extinction; wherever possible this has been noted.

The heading for each species gives the most common English name and where alternatives occur these are given in the text. The order, family and specific Latin name are given in the captions as well as the length (beak to tail), breeding habits and additional information about related species. The distribution of each species is included and refers to the areas in which the bird breeds. In the case of migratory birds the regions in which they winter are also given.

ADELIE PENGUIN

Order Sphenisciformes, comprising 17 species. The penguin is the most marine of all birds. Its wings have become hardened and transformed into flippers, and the webbed feet act as paddles. They swim at speeds of up to 40 km (25 miles) per hour. Diet consists of fish, crustaceans and occasionally squid caught just below the surface of the water. Adelie Penguin (*Pygoscelis adeliae*) is 70 cm (28 in) long. Eggs are incubated for 33–38 days by the males who do not enter the water to feed during this period. Young are silvery-white, and later a dirty brown. At four weeks they are left to themselves for long periods and gather in nurseries of up to 200 individuals. The Galapagos Penguin (*Spheniscus mendiculus*), is an endangered species.

A thick, dense, oiled coat protects penguins from freezing temperatures but also enables them to move smoothly through water. A layer of fat beneath the skin gives further insulation and provides a food reserve when the birds come ashore to breed. Penguins have been flightless for about five million years, but they are extremely agile swimmers and leap from the water like porpoises. The Adelie Penguin nests on the shores of Antarctica and adjacent islands. In common with other penguins it gathers in huge colonies often numbering over one million individuals. A white ring of skin circles the eyes and the beak is short and partially covered with feathers. The Adelie also has a long tail of feathers which trails behind it as it walks or runs. Most penguins are monogamous and lay two to three eggs in a nest fashioned from a rough pile of stones. The male Adelie incubates the eggs.

OSTRICH

Ostriches, the largest of all living birds, belong to the group of flightless birds known as ratites, or running birds. This group also includes the rheas of South America, the Australian emus and the cassowaries of Australia and New Guinea. Although unrelated, they all lack a carina or keel, the bone found in all other species which supports the strong muscles needed for flight. These muscles are weak and reduced in ratites. Long, powerful legs compensate for the inability to fly: ostriches can run at a speed of up to 70 kilometres (43 miles) per hour and can jump 1.5 metres (5 feet). Gregarious birds, they are also protected by living in huge flocks, often in the company of zebras and antelopes, and by their excellent vision. Ostriches sit on the ground to sleep, normally with the neck erect. When, for short periods, they fall into a very deep sleep the neck is stretched before them on the ground.

Order Struthioniformes, ostriches. The Ostrich (*Struthio camelus*) is 2.5m (8ft) tall. Once widespread throughout Africa and parts of Asia, now only one species, two races, is found. The Northern Ostrich has a bare patch on its head; males have scarlet skin, and females are bright yellow. Southern Ostrich is slightly smaller. The skin is light grey in the females, and dark blue in the males. Both races inhabit deserts and savannahs. Its lumbering gait and the fact that it can go without water for several days have given it the name Camel-bird. Males are polygamous. Nest, made by males, is a scrape. Eggs, 10–12, are incubated by the female during the day and by the male at night.

GREAT CRESTED GREBE

Order Podicipediformes, comprising 18 species distributed throughout the world. Although they are water birds, the toes are not completely webbed. Each has a fringed, fin-like lobe, a characteristic shared with the Coot (page 151). The diet is mainly fish, but crustaceans, insects and their larvae are also taken. Many also have the curious habit of eating their own feathers. The Great Crested Grebe (*Podiceps cristatus*) is 48cm (18½ in) long, and is found in Eurasia, Africa, New Zealand, Australia and Tasmania. Eggs, 3–4 are laid from April to July in Europe. North American species include the Pied-billed Grebe (*Podilymbus podiceps*).

The grebes are water birds that were once grouped with the divers and loons but are now classed as a separate order. The Great Crested Grebe, one of the largest of the group, is easily distinguished in spring and summer by two feathery grey tufts on the top of its head and a ruff round the top of the neck. Both male and female display their plumage during courtship and participate in an elaborate courtship ritual which includes much head-shaking and loud cries. The performance culminates in a 'dance' in which the birds rush rapidly towards each other, rise from the water and, touching breast to breast, exchange nesting materials. The nest is made from floating vegetation and is attached to reeds and other plants at the edge of the lake or pond. Grebes are cautious and extremely protective parents. On leaving the nest they conceal their eggs by covering them with leaves. The young are carried on their parents' backs.

Order Podicipediformes. Family Podicepididae. The Little Grebe or Dabchick (*Podiceps ruficollis*), 27cm (10½in) long, is found in Europe, Asia and Africa. Its winter plumage is light brown with pale underparts. Mating call is a loud trill and both male and female take part in courtship songs. Unlike the Great Crested Grebe, which often nests in colonies, Dabchicks are strongly territorial and defend the nest against intruders. The nest is a floating mass of vegetation. Eggs (5–7), bluish-white at first, are later covered with dark stains. Two broods are raised each year: the first as early as March and the second as late as September. A related species found in Australia is the Hoary-headed Grebe (*P. poliocephalus*).

LITTLE GREBE

The Little Grebe or Dabchick, one of the smallest of the Old World species, breeds on inland waters and streams. It builds a solitary, well-hidden nest among reeds and other water plants. The breeding season lasts from March to September, when at least two and sometimes three broods are raised, each consisting of three to six striped chicks. As soon as the chicks hatch they take refuge among the parents' back feathers or under their wings. The Little Grebe lacks a crest but during spring and summer the front of the neck is chestnut red and the corners of the bill yellow-green. Like other grebes the Dabchick is more at home on water than on land and is an expert diver, submerging for about 15 seconds to search for food or escape from danger. All grebes have short tails and short wings, and most are weak fliers. Some species, however, such as the Little Grebe and the Great Crested Grebe, migrate for the winter.

ALBATROSS

The albatross has been associated with ships and sailors for hundreds of years. Seamen sometimes call it the mollymawk or mollyhawk, from the dutch *mollemoks*, meaning 'stupid gull'. Yet albatrosses are far from stupid and in fact are affectionately regarded by seamen as a symbol of good luck, especially when they follow in the wake of a ship for days on end. The air is their natural element; most of their life is spent gliding on thermals and the strong winds that blow freely throughout the southern hemisphere. Without these winds the albatross is virtually helpless, being an excellent glider but a poor flier, and is forced to settle on the water in still conditions. The Wandering Albatross, the largest of all living flying birds, is also one of the most impressive marine birds. In common with some other species it comes ashore only during the breeding season. This usually occurs once every two years.

Order Procellariiformes, the tubinares or tube-nosed birds, which also includes shearwaters and fulmars (page 119). All are sea birds with a hooked, horny-plated bill, tubular nostrils, webbed feet and thick plumage. Diet includes fish and squid. Some also feed on waste from ships and on whale dung. The Wandering Albatross (*Diomedea exulans*) has a wing span of 3m (9½ft). It inhabits waters of the southern oceans and nests on sub-Antarctic islands south of New Zealand and in the south Indian and Atlantic Oceans. The Royal Albatross (*Diomedea epomophora*), which is very similar, nests on Auckland, Chatham and Campbell Islands and near Dunedin on South Island. There are eleven further species of albatross in the world.

FULMAR

Order Procellariiformes. Family Procellaridae – shearwaters and fulmars. Northern Fulmar (*Fulmarus glacialis*), 48 cm (19 in) long, is found on sub-arctic shores of the Atlantic and Pacific Oceans. It migrates south in winter to coasts of Japan, France, eastern Canada and California. Silver-grey Petrel or Southern Fulmar (*Fulmarus glacialoides*), 46 cm (18 in) long is found in Antarctic seas and southern oceans. Some ornithologists believe both birds may be one species. Plumage is dark smoky-grey or greyish-blue. The more northerly the species the darker they tend to be. They feed on plankton, carrion and fish offal. The breeding areas of these birds have increased in the past 200 years owing to the spread of commercial fishing.

The word fulmar means foul-gull, for these birds have a habit of spitting a nasty-smelling oil at intruders to the nesting site. Closely related to petrels and shearwaters, fulmars are one of the most numerous of sea birds, their numbers having increased rapidly during the past 200 years. Northern Fulmars are also called mollies by sailors and, like the albatrosses (page 118) and storm petrels, are believed to be reincarnations of people who drowned at sea. In size and plumage they resemble the Herring Gull (page 163) but have shorter necks and dark-ringed eyes. When Fulmars are not nesting they cruise over the ocean, sometimes flying low over the surface or landing on the water. They are extremely clumsy on land and fly directly to the open nests which are generally made on steep cliff faces. The nest is usually a niche in the rock or a scrape on the ground. No nesting materials are used, but stones are used to decorate the edge.

PELICAN

Pelicans are a group of aquatic birds that are differentiated from all others by their four webbed toes. Another distinguishing characteristic is the throat pouch, or gular, which is particularly well-developed in the pelicans. The pouch hangs down from the lower mandible and is used like a fishing net to capture fish and other animals; the upper mandible forms a lid to the pouch which can hold up to 13.5 litres (3 gallons) of water. Contrary to popular belief, the pouch is not used to store food. Pelicans move clumsily on land but hollow bones and air sacs under the skin lighten their weight and help to make them strong and accomplished fliers. They often fly in straight lines like ducks and geese, with their wings making a rhythmic, syncopated beat. Many pelicans immerse themselves underwater to hunt for food but because of the air sacs most cannot dive to any great depth except for the American Brown Pelican.

Order Pelecaniformes. Family Pelecanidae. Seven species are found throughout the world with the exception of Antarctica. Eggs 3–4, are white. The nest is a rough-and-ready structure made in a hollow near water, surrounded by tall plants. Young hatch naked and fly at about 2 months. The Eurasian White Pelican (*Pelecanus onocrotalus*), 1.7m (67in) long, is found in eastern Europe, central Asia and southern Africa. The one species in Australia and New Guinea is the Australian Pelican (*P. conspicillatus*). The American White Pelican (*P. erythrorhynchos*) is found from western to central California, North Dakota and Mexico and Florida. Brown Pelican (*P. occidentalis*) is found in southern States and South America. The average wing span is 2.5m (2½ft).

SHAG

The Shag, a large bird with a dark, glossy green coat, is known in some parts of Europe as the Green Cormorant. Shags and Cormorants are very closely related; all of them have a long hooked beak, webbed feet and a short wedge-shaped tail. In Europe the Shag is sometimes confused with the Great Cormorant, a larger bird with a white patch on the chin and, during the breeding season, a white patch on the thighs. Shags lack the white markings and have a crest on top of the head in spring. They are more aquatic than the cormorants and spend much of their time swimming low in the water along rocky coastal shores. They are poor fliers but can swim underwater with great agility and speed, paddling with their feet together and their wings half open; the tail is used as a rudder. Like their relatives they dive for food, entering and leaving the water in an arc. Dives may be to 10 metres (32 feet) deep lasting for a minute.

Order Pelecaniformes. Family Phalacrocoracidae – Shags and Cormorants – comprising 30 species, found in most parts of the world. Shag (*Phalacrocorax aristotelis*), 78 cm (30½ in) long, breeds along northern European and Mediterranean coasts. They sometimes migrate south in winter. They lay 2–3 pale blue eggs in a nest of seaweed built on rocky ledges. Food consists mainly of fish and some crustaceans and molluscs. Great or Common Cormorant (*P. carbo*), 93 cm (36 in) long, is the largest and most widely distributed of the family. It is found in Europe, Asia, Africa, north-eastern Canada, Australia and New Zealand. A similar species, the Double-crested Cormorant (*P. auritus*) is found from Alaska to Central America.

GREY HERON

Order Ciconiiformes. Family Ardeidae, comprising 62 species of herons, egrets and bitterns. Grey or Common Heron (*Ardea cinerea*), 91cm (3ft) high with a wing span of 1.8m (6ft), is found in Europe, central and southern Asia, Africa and Madagascar. The nesting site is chosen by the male before he begins courtship, which includes the presentation of nesting materials to the female. The nest is made of dead sticks and reeds and is lined with grass and other plants. Heronries are found in trees, especially in deciduous woods, and sometimes in coniferous woods. They are also found in marshes or bushes. Eggs, 4–6, are greenish-blue. Diet includes fish, small mammals and aquatic animals. New World relative, Great Blue Heron (*A. herodias*) is also found.

The Grey Heron, a large wading bird of rivers and lakes, has the three partially webbed toes, long beak, neck and legs typical of the order Ciconiiformes – the storks, ibises and their relatives. These characteristics are adaptations for hunting in shallow waters. Standing completely motionless, the heron watches for the movement of fish, frogs and other small water creatures. Having sighted its prey, it suddenly darts out its flexible neck and grasps the animal in its strong beak. The Grey Heron is a shy, timid bird that nests in large colonies, usually in tall trees or among reeds. Like all herons it flies with its legs stretched backwards and its neck bent in an S shape. Interesting features of this bird, shared with most other members of the order, are the 'comb' on the middle toe and the presence of powderdown under the breast which frays into powder and is used to clean soiled feathers.

EGRET

Smaller than herons (page 122), egrets are slender, beautiful birds with brilliant white feathers. They are now a protected species, but during the 18th and 19th centuries thousands of them were killed for their elegant breeding plumage, which was used to adorn women's clothing. The long plumes appear on the shoulders and breast, with a lengthened crest at the back of the head. Egrets were once worth twice their weight in gold. The slaughter of four egrets produced only one ounce of the valuable feathers. The most common species in the Old World is the Little Egret. It breeds in large colonies on river marshes and swamps and builds its nest in trees and among reeds. Like the other species it stalks its prey then, like the heron, grabs the animal with lightning speed. The Great Egret is rare in Europe but occurs quite frequently in most other parts of the world. Three North American species are now extinct.

Order Ciconiiformes. Family Ardeidae. Little Egret (*Egretta garzetta*), 65cm (25½in) long, is found in Europe, Asia and North Africa. Greatest concentrations are found in the Camargue (France) and Coto Doñana (Spain). It inhabits river marshes and swamps and nests on dry and wet ground or in trees such as poplar, elm, and ash. The nest is solitary or alongside nests of the Night and Purple Herons. Eggs, 3–6, are greenish-blue. The diet is extremely varied including small fish, insects, amphibians, small mammals, snakes, worms, lizards and snails. Great Egret (*E. alba*) 102cm (40in) long, is found in Europe (along the Danube and in southern Russia), Australia, Africa and Asia. New World species, Snowy Egret (*E. thula*), is 61cm (24in) long.

MARABOU STORK

The family Ciconiidae contains 17 species of storks and jabirus, long-legged birds with short, partially webbed toes and strong wings. They are all excellent fliers, flying with out-stretched wings and, taking advantage of strong air currents, soaring high above the ground. The adults are almost mute and can produce only a hissing sound; they communicate with others and their young by rattling their beaks. The African storks are the giants of the family, measuring over 1.5 metres (5 feet) in height. The bill is usually covered with a cere, or thick skin, at the base, sometimes in the shape of a saddle.

Most storks are elegant birds, admired for their beauty and grace. The Marabou Stork of Africa and Asia is a remarkable exception. Its name derives from the French word *marabou*, meaning a holy man or, in this case, an ugly deformed person. The Marabou has a long sharp beak which is used for piercing, a naked pink head and neck and, as in most other species, a long fleshy pouch under the throat. The pouch is not a crop but is composed of connective tissue. It is used as a pad on which the Marabou rests its beak. This stork struts about on its long legs with a military gait, a characteristic that has earned it the alternative name of Adjutant Stork. Its long tail feathers were once collected and used to decorate women's clothes. The Marabou is primarily a scavenger and, in spite of its disagreeable appearance, it is generally welcomed near human habitations for its function as a disposer of rubbish. It also hunts live prey and, although it rarely ventures into water, during periods of drought it will wade in shallow waters to probe in the mud. On the African savannahs it competes with the vultures (page 139) for food, plunging its naked head inside a carcass and piercing the abdominal walls to tear out the intestines. The Marabou breeds in colonies in trees or on rocks. The eggs are laid at the end of the rainy season so that rearing of the young coincides with the dry season when carrion is plentiful and large numbers of aquatic animals are accessible in dried pools and waterways. Only a small proportion of sexually mature birds breed in any one season, the rearing of the young taking up a considerable part of the year. Three species are distinguished: the Greater and Lesser Marabous of Asia and the Marabou of Africa.

Order Ciconiiformes. Family Ciconiidae – storks and jabirus. Marabou Stork (*Leptoptilos crumeniferus*) is 150 cm (60 in) long with a wing span of 3 metres (9ft). The pouch is 30 cm (12 in) long. Found in Africa south of the Sahara, it nests in colonies in the same places as herons and other storks. Eggs, 2–3, white, are incubated for 30 days; in 116 days the young fledge and after 130 days they leave the nest. The Marabou Stork feeds on fish, small mammals, carrion and refuse, and eats about 720 grams (1½ lb) of food each day. Greater Marabou (*L. dubius*) is found from central India to Borneo. Lesser Marabou has a wing span of 3.2m (10½ft) and is found from central India to Borneo and Java.

WHITE STORK

Order Ciconiiformes. Family Ciconiidae. White Stork (*Ciconia ciconia*) is 1.2m (4ft) tall with a wing span up to 2.1m (7ft). European breeding grounds are in Alsace, Holland, Germany and Scandinavia. The nest may reach enormous proportions, especially the old ones which can be up to 2.4m (8ft) in diameter and weigh 45kg (100lb). They are made from a variety of materials including lumps of earth, straw and reeds. Four white eggs are incubated for about one month. North American species, the Wood Stork (*Mycteria americana*), is similar to the White Stork but has a black tail and pink legs. Formerly misnamed the Wood Ibis, it is found from the Florida swamps south to Argentina.

In many parts of the world the White Stork is regarded as a symbol of good luck and fertility, and its arrival at its European breeding grounds in February and March is usually taken as the first sign of spring. These storks are monogamous and pair for life, returning to the same nesting site year after year. The nests may be built in trees (usually in colonies), but these sociable and partially domesticated birds will often nest on roof tops and chimneys, sometimes encouraged by local people who place a basket or wagon wheel on their roof as a base for the nest. The location of the breeding site is largely determined by the availability of water where the bird can hunt for frogs, molluscs and crustaceans. It also frequents fields, searching for small rodents and insects. The fledglings are taught to hunt when they are about two months old. After a few foraging flights the young leave the parents.

Order Ciconiiformes. Family Ciconiidae. Saddle-billed Stork (*Ephippiorhynchus senegalensis*), 1.6 m (5½ ft) long with a wing span of 3 m (10 ft), is found in tropical Africa from Ethiopia and Senegal to South Africa. It lives near rivers and lakes. The nest, a substantial structure made of sticks is usually built in trees and bushes but sometimes on cliff edges. Eggs, 1–4, are dull white and pitted. It diets on fish, frogs, small mammals, birds and some insects. The Jabiru Stork (*Jabiru mycteria*) is the third largest of the storks and the largest New World species. It is 1.2 m (4ft) long, and is found from Florida south through Mexico to Argentina in lakes and swamps. Black-necked Stork (*Xenorhynchus asiaticus*) is found from India to north-western Australia.

SADDLE-BILLED STORK

The Saddle-billed Stork is an enormous bird of tropical Africa. It is easily distinguished from all other storks by its splendidly coloured red, yellow and black bill. The base of the bill is covered with thick skin (cere) in the shape of a saddle. The head, wing coverts, neck and tail are a brilliant black and the remaining parts, including the flight feathers, are pure white. Like all storks the Saddle-billed Stork is a powerful flier but it flies with its head and legs lower than its body. It does not gather in large groups but is commonly seen in pairs or small troops near swamps and along the edges of lakes. It feeds by slowly stalking its prey, then stabbing it with the bill. Breeding occurs in isolated pairs. Courtship display is accompanied by loud bill-clacking. The largest of the storks, the Saddle-billed is closely related to the Black-necked Stork of Asia and Australia, which is almost as large and looks very similar but lacks the saddle.

WHITE SPOONBILL

The long, flattened beak of the Spoonbill, widening at the tip into a spatula, is specially adapted for filtering food from shallow waters. Usually feeding at night and at low tide, the Spoonbill moves its head from side to side with the bill half-open and partly submerged in the water. In Europe the White Spoonbill nests in Holland, the lands along the Danube and in the south of Spain, but it is sometimes seen as a visitor to other countries including England. Its breeding grounds are marshy places which are often shared with bitterns and herons. The nests are built in bushes close to the water's edge or among reeds on small islands. Both male and female take an active part in incubating the eggs and rearing the young. The Roseate Spoonbill of the New World is a close relative and apart from its rose-pink colouring is similar in habit and appearance. Recently both species have begun to decrease in numbers.

Order Ciconiiformes. Family Threskiornithidae. Spoonbills and ibises, of which there are six species of Spoonbill. White Spoonbill (*Platalea leucorodia*), 90cm (36in) long is found in Europe and Asia. It inhabits wet, marshy areas, coastal shores and estuaries. The nest is made of twigs, sticks and reeds. Eggs, 3–5, are white with reddish-brown spots. It diets on fish, insects, crustaceans, worms and frogs. Roseate Spoonbill (*P. ajaja*) is found from Florida south to Argentina. Other species include the African Spoonbill, a white bird with bright pink legs, and the Yellow-billed Spoonbill of Australia.

GREATER OR ROSEATE FLAMINGO

Greater Flamingo (*Phoenicopterus ruber*) is 127cm (51in) long. Both male and female have black tips to the wings and bill. It is found in the Mediterranean regions of Europe, in Asia, parts of North and South America and in Africa, especially on the shores of Lake Nakuru in Kenya, where thousands are found. Nest is made of mud, with a hollowed out centre. The bird lays 1–2 white eggs. Classification of the flamingo is still under debate. Because of its wading habits it is often grouped with the storks and herons, or sometimes with the ducks and geese, with which it shares a similar beak structure. Other zoologists look on it as a separate order because of the complete webbing of the three front toes.

These beautiful and unusual birds are found throughout the temperate and tropical regions of the world. Whether in the air or on land they throng together in vast numbers, creating a splash of colour against the sky or water. At rest they fold their heads back on their bodies, with one leg tucked under the belly; during flight the legs are stretched out behind and the long, graceful neck is stretched forward. Flamingos have a unique, highly specialized bill which bends downward in the centre. The Greater Flamingo gathers crustaceans, insect larvae and other small animals from the mud by twisting its neck and holding the bill upside down. It then uses its thick tongue and tooth-edged bill to sieve the food. The colour of the flamingo's plumage is caused by the presence in its diet of a substance called carotinoid. When carotinoid is absent the bird's feathers become a dirty white, as happens with captive flamingos.

MUTE SWAN

Swans are often regarded as symbols of purity and many romantic legends surround these graceful birds, yet they are strongly territorial and extremely aggressive in defence of their young and their nests. The Mute Swan, native to Europe and Asia, is a typical member of the group; pairs are monogamous and mate for life, often occupying the same nesting site during their long life span. The cob (male) is remarkably solicitous of the pen (female), caressing her with his bill, gathering nesting materials and feeding her during the incubation period. Both parents care for the cygnets when they hatch, often carrying them on their backs during the first few days. In spite of their name, Mute Swans are not voiceless – intruders are frequently warned off by loud hissing noises accompanied by a wild beating of the wings. The breeding season is followed by two moults; during one, all the flight feathers are lost and replaced.

Order Anseriformes, ducks, geese and swans. Mute Swan (*Cygnus olor*), 1.5 metres (5ft) long, has a life span of 30 to 40 years. Wild populations are found in northern Europe and parts of central and eastern Asia. Eggs, 5–9, are grey-green. The offspring of captive Mute Swans now breed in the wild along the coasts of the mid-Atlantic states of the United States. A closely related species, the Black Swan (*C. atratus*), inhabits southern Australia. The Trumpeter Swan (*C. buccinator*) of north-western North America is the rarest of the swans. Once almost slaughtered to extinction for its swanskin, it is now a protected species and numbers are increasing.

MALLARD

Order Anseriformes. Family Anatidae, comprising 43 genera. Mallard (*Anas platyrhynchos*), 58 cm (23 in) long, is found across Europe and Asia except for northerly and southerly extremes, in most of North America as far north as the Arctic Circle, and in parts of north-west Africa. It inhabits ponds, marshes and the edges of lakes and waterways. The nest of stems, leaves and grass is made on the ground among thick cover or in branches of bushes or on buildings. Mallards sometimes take over the abandoned nests of other birds. Eggs, 5–16, are usually green but colour varies to brown and grey. Its call, the familiar *quack*, is delivered only by the female; the call of young is a high-pitched sound.

The Mallard or Wild Duck is the ancestor of many species of domestic ducks. It belongs to the group of dabbling ducks, birds that forage for food by dipping their head and neck below the surface of the water or by upending their bodies. The Mallard is a voracious feeder and eats most water plants and animals, from seeds and leaves to small fish. Mating takes place on the water with the female completely submerged. Both duck and drake choose the nesting site but after that the drake takes little interest in the brood. The duck lines the nest with down from her breast, incubates the eggs alone and cares for the young. Ducklings leave the nest as soon as they hatch, and swim after their mother. They make their first flight at about two months. During the moult the drake loses its bright, conspicuous plumage and takes on the drab colours of the female (eclipse plumage). At this time they remain out of sight among the reeds.

WOOD DUCK

Native to North America, the Wood Duck is a splendidly coloured species now found as an ornamental bird in many parts of the world. As with most ducks it is the drake that exhibits the colourful plumage, the duck being much duller. One of the few species of perching ducks, the Wood Duck makes its nest in tree holes or in nest boxes. The young do not fledge in the normal manner. They are equipped with sharp claws and climb up the side of the nest and then drop to the ground. The Mandarin, which is closely related to the Wood Duck, has been exported to many countries from its native Asia. The drake has a glossy, dark-green head with white markings, the sides and neck are purplish-green with a metallic sheen and the breast is white. Breeding pairs are extremely affectionate and in China the Mandarin is a symbol of conjugal fidelity.

Order Anseriformes. Family Anatidae. Wood Duck (*Aix sponsa*), 48 cm (19 in) long, is native to the south-eastern United States and winters in south Mexico. Feral species are found in other regions of North America. The Wood Duck inhabits lakes and inland streams. Nests are made in tree hollows and are lined with down from the duck's breast. (Eggs: 1–15.) Mandarin Duck (*Aix galericulata*) is a native of Asia and the Far East (Japan and Formosa). It has been successfully introduced to southern England where it is found wild. Both species are often seen in parks and gardens and are highly valued by bird fanciers.

SHOVELER

The Shoveler gets its name from the unusual, long flattened beak which is specially adapted for feeding in shallow water. The comb-like lamellae (horny, toothed plates) found at the sides of all ducks' bills are well-developed and noticeable in this bird, being fine and densely distributed along the edges of the bill. Shovelers feed by moving the beak from side to side, using the lamellae to separate microscopic animals from aquatic plants, mud and water. This manner of feeding is similar to that of the flamingos (page 129). Because of its heavy bill, the Shoveler is weighted down at the front when it sits on the water. It is also distinguished by its habit of swimming in a keel formation when in a group. The female is typically more sombre in colour than the drake but she has a similar build and stature.

Order Anseriformes. Family Anatidae. The Shoveler (*Anas clypeata*), 51cm (20in) long, is found in almost all parts of Europe, in northern Asia and North America. In North America it is known as the Northern Shoveler and occurs in northerly regions of the eastern and western United States and in Northern Canada. The Shoveler migrates south in winter. It inhabits areas of shallow water, particularly swamps and marshes, but is also found on moors and meadows in the breeding season, not necessarily near water. The nest is made of plant materials and lined with grass and down. It is usually placed among plants or in a hollow. Eggs, 7–12, greenish-grey, are incubated for about 25 days. The young leave the nest shortly after hatching.

COMMON TEAL

Teal are the smallest of the European and North American ducks, being about the size of a pigeon and half the size of the Mallard. The Common Teal of Europe is an extremely lively bird and a strong flier with an erratic flight pattern. Like all dabbling ducks it leaves the water suddenly without any preliminary paddling, but instead of giving the usual head-shake indicative of intended flight it nervously bobs its head up and down. When disturbed, Common Teal take off in a group, rising almost vertically into the air. The Common Teal is a retiring bird. It breeds in remote areas or in places with abundant vegetation so that the nest will be well hidden from intruders. The nest is usually lined with several layers of down which are added to throughout the brooding period. The duck also covers the eggs with down when she leaves the nest.

Order Anseriformes. Family Anatidae. Common Teal (*Anas crecca*), 35cm (14in) long, is found in most of Europe and northern and central Asia. It occasionally wanders to North America. It inhabits shallow pools and breeds on moorland, in woods and ditches. The nest is made from plant materials and is lined with down. Eggs, 5–10, are pale cream with a greenish tinge. Young leave the nest shortly after hatching. Blue-winged Teal (*A. discors*), 36cm (14in) long, is found in eastern and western parts of North America. Green-winged Teal (*A. carolinensis*) is found in western North America and migrates to eastern regions in winter. Baikal Teal (*A. formosa*) is found in north-east Asia and winters in China and Japan.

GARGANEY

The Garganey is sometimes erroneously grouped with the teal but it is a slightly larger bird with a more direct flight pattern. It is more closely related to the Shoveler (page 133) than the teal. Sometimes it is called the Cricket Teal. The drake has a curious cracking call during the breeding season. Both sexes have a white stripe above the eye but in the duck the stripe is not as bold. Many of the dabbling ducks have a well-defined speculum (a patch of bright feathers on the wings), the colouring of which is often used to identify species. In the Garganey the speculum is relatively ill-defined. As with most ducks the female Garganey is less conspicuous than the male, and is brownish with white flecks. The Garganey has a limited range and is one of the rarest of the ducks. It has a preference for still or slow-moving waters surrounded with rich vegetation.

Order Anseriformes. Family Anatidae. Garganey (*Anas querquedula*), 41cm (16in) long, is found in temperate regions of Europe and Asia, usually in regions with abundant vegetation. The nest is made from local plants and is lined with down. The nesting site is on isolated stretches of water or meadows, marshes or moors. Eggs, 8-11, creamy brown are incubated for about 23 days. The young closely resemble the young of the Mallard but have two dark stripes on the sides of the head. They vacate the nest shortly after hatching.

SHELDUCK

One of the most conspicuously coloured of the Anatidae, the Shelduck is easily recognized by its upright stance and bright breeding plumage (the world 'sheld' means variegated or piebald). A bird of coastal waters with a scattered distribution throughout Europe, in habit and appearance it is a curious blend of the duck and the goose. Duck-like features include the shape of the head and biannual moults. As in geese there is no distinction between the plumage of the two sexes. The female is slightly smaller than the male and lacks the knob at the base of the beak. During June and July thousands of Shelduck from north-western Europe gather on the sands at Knechsand, on the North Sea coast of Germany, to moult. Only females with young remain behind at the breeding site. The eclipse plumage differs only slightly from the breeding plumage; colours are duller and white patches appear on face and throat.

Order Anseriformes. Family Anatidae. Shelduck (*Tadorna tadorna*), 60 cm (23$\frac{1}{2}$ in) long, is found along coasts of Spain, the British Isles, Scandinavia and through the temperate regions of Eurasia to Manchuria. In September it migrates to winter quarters in North Africa and southern Asia. It prefers land to water although it sometimes swims. The nest is made in sandy burrows, frequently in rabbit holes, but usually on salt marshes or sand dunes. Nests made above ground are hidden under thick vegetation. The female lines the nest with down. Eggs, 7–12, are incubated by the female. Drake, like the gander, takes no part in brooding but usually stays close to the nest and guards the female. The Shelduck feeds while wading in shallow water.

Order Anseriformes. Family Anatidae. The tribe Somateriini comprises four species. Eider (*Somateria mollissima*), 58 cm (23 in) long, is distributed throughout the northerly regions of the northern hemisphere. It inhabits barren rocky and sandy shores and nests in large colonies, usually near coasts but sometimes near inland freshwater lakes and rivers. The nest is constructed by both parents from grass and seaweed and is made on the ground. Thereafter the male takes no interest in the eggs or the young. Eggs, 4–6, are glossy greenish-grey. Normally there are three broods a year. Diets on molluscs and crabs; the hard shells are broken with the strong beak. Food is often collected while diving.

EIDER

Eiders are sea ducks found on northern coasts and islands throughout the northern hemisphere. The Common Eider produces an abundance of soft, white down during the breeding season. In many regions the down is collected and sold commercially as the well-known eiderdown. Collection is usually made by fencing-in breeding areas and gathering the down from the nests after the first and second clutches have been laid. Eiders are short-necked, heavy birds with wedge-shaped heads and sturdy bills. The breeding plumage of the drake is a striking black and white; the sides of the head and nape are pale green and the breast is often tinged with pink. The duck's plumage is brownish with dark brown bars. Following the first moult in July the male becomes almost completely black, with small patches of white on the wing. The eclipse plumage lasts until the second moult in December.

GREYLAG GOOSE

Wild Geese are inhabitants of the northern hemisphere. They breed in the cold tundra regions and migrate south in the winter to the Mediterranean coast and Africa in the Old World, and to the southern United States and Mexico in the New World. The Greylag, or Wild Goose, is the typical representative and best-known European member of the group; the legs and neck are long, although shorter than in swans, and the thick beak ends in a nail. Overall size is mid-way between that of the swan and the duck. The ancestor of most farmyard geese, the Greylag is noisy and aggressive, particularly in the breeding season when it will attack invaders to the nesting site. Mating takes place on the water after mutual display. In other respects the Greylag is more terrestrial than other wildfowl and much of its time is occupied by grazing. The domestication of this bird has been traced back to ancient Egypt around 2000 B.C.

Order Anseriformes. Family Anatidae. Fourteen species of geese make up two genera, the *Anser* and the *Branta*. The plumage of both sexes of geese is alike. Greylag Goose (*Anser anser*), 90 cm (3 ft) long, is found in the wild in parts of western Russia, Germany, Poland, Iceland, the Faeroe Islands and north-west Scotland. The Greylag Goose breeds in solitary pairs or in groups It inhabits tundra, moors, reed beds and marshes, and feeds on a vegetarian diet of grains, grasses and tubers. The nest is made of twigs and stems of local plants. Eggs, 4–9, are creamy white. The Greylag vocalizes by loud honking or threatening hisses. It is often bred commercially and force-fed with grain to enlarge the liver.

VULTURES

Vultures occupy two large groups among the birds of prey: the New World and the Old World vultures. Although the groups are separated geographically and have certain anatomical differences, both fulfil an important ecological role by disposing of animal remains and faecal matter. To many people the Old World vultures are particularly repellent in appearance but they are perfectly adapted to their role. The head and long neck, naked and usually coloured bright red, enable the bird to reach far into animal cavities; the sharp, hooked beak is designed for shearing and dismembering. Vultures glide for hours on thermals, searching for carrion, then suddenly plunge to the ground and approach the carcass with neck thrust forward and wings half open. In Africa, several different species may feed on the same corpse at the same time, in company with the Marabou Stork (page 124).

Order Falconiformes, birds of prey. The family Accipitridae contains nine sub-families, including the Old World vultures. With a wing span of 2 metres (6½ft), the Egyptian Black Vulture (*Aegypius monachus*) is one of the largest. The plumage is dark brown with a collar of feathers at the base of the neck. It is found in central Asia, North Africa and countries bordering on the Mediterranean. The family Cathartidae contains all the New World vultures (commonly called buzzards in North America). The best-known and largest species is the Andean Condor (*Vultur gryphus*), with a wing span of 3.6 metres (12ft). It is found in its largest numbers throughout Peru and Chile. The plumage is shiny black with white bands on the wings and a white collar at the base of the neck

GOLDEN EAGLE

Order Falconiformes. Family Accipitridae. Golden Eagle (*Aquila chrysaetos*). The female can grow up to 83 cm (33 in) long with a wing span of 1.8 metres (6–7 ft); the male is slightly smaller. Found in mountainous regions of Europe, Asia and North America, it nests on ledges on very steep rocky cliffs and high in trees. In remote places it may also nest on the ground. The nest is made of branches and earth and is decorated with foliage. Eggs, 2–3, are white, some with reddish or grey markings. The young remain in the nest for up to 80 days. Golden Eagles feed on hares, rabbits, carrion, grouse and other birds, and sometimes small lambs.

One of the most impressive of all birds, the Golden Eagle was once widely distributed throughout the northern hemisphere; today it is a rarely-seen species found in relatively small numbers in the mountainous regions of Europe, North America and Asia. In Britain its range is now restricted to the Scottish Highlands. The word majestic is often used to describe this bird which has a confident perching stance and a soaring flight, often reaching 500 metres (1692 feet) then diving towards the ground with its wings half closed. Seen from a distance it may be confused with the buzzard (page 142) which also spreads its flight feathers when it is soaring. In common with other eagles, the female is slightly larger than the male although both sexes have the same brownish plumage with yellow tinges on the head, neck and shoulders.

HEN HARRIER

Order Falconiformes. Family Accipitridae comprising 17 species. Hen Harrier (*Circus cyaneus*) is known in North America as the Marsh Hawk. The hen is 45 cm (18 in) long. It is found in most parts of Europe, North America, Asia and North Africa and prefers flat terrain such as heaths and moors, and is sometimes found in young forests. The Hen Harrier nests among field crops or other suitable ground cover. The nest is made in a hollow of local plants such as heather, and lined with grass. Eggs, 4–5, are chalky white. Young remain in the nest for about a month. There are some migratory movements southwards in winter. The female Hen Harrier is very similar to the hen of Montagu's Harrier (*C. pygargus*), 42 cm (17 in) long.

Harriers are birds of the northern hemisphere. They are sometimes called Marsh Hawks but they comprise a separate subfamily and differ from true hawks in several important ways. All harriers nest on the ground and the sexes are markedly different in their plumage. The English name comes from the manner in which they harry or harass their prey by flying back and forth very low over the ground. Smaller and more slender than hawks, harriers are also distinguished by the markings round their eyes which give them an owlish appearance, by their long tail feathers and by the ruff round the front part of the head. The female Hen Harrier (illustrated here) is distinguished by its white rump. The male, which is smaller than the female, is grey and white with black wing tips.

COMMON BUZZARD

Buzzard is a name commonly applied to several different birds of prey, but strictly speaking it refers to the 26 species of the genus *Buteo*. In Europe the Common Buzzard is the most numerous of the raptors and is often seen gliding over fields and moors. A wary bird, it usually secretes itself in bushes and trees bordering open ground from where it watches for small animals. Carrion and occasionally vegetable matter are also eaten. Closely related to the eagle (page 140), it has the same large build and rounded wing tips but can be distinguished from the eagle and other birds of prey by the short beak which is curved from the base downwards, and by the absence of feathers on the lower leg. In North America the buzzards are termed hawks; the word buzzard is used as a common term for the New World vultures (page 139).

Order Falconiformes. Family Accipitridae. Subfamily Buteoninae, true buzzards. Common Buzzard (*Buteo buteo*), 54cm (21in) long, is found in most of Europe, Asia and Africa. Some northerly birds migrate further south in winter. The Common Buzzard inhabits woods, copses and occasionally cliffs; and eats a variety of prey: insects, amphibians, mice, moles and reptiles. The nest is made of twigs and lined with moss and fur. It is built high off the ground in trees. Eggs, 2–3, are white, and some have reddish-brown to purple markings. There are several New World species including the Red-tailed Hawk (*Buteo jamaicensis*), 56cm (22in) long, which is found from Alaska to central America. Plumage colour is extremely variable in all species of buzzard.

Order Falconiformes. Family Accipitridae. Black Kite (*Milvus migrans*), 56 cm (22 in) long, is found in Eurasia, Africa and Australia. In winter it migrates. It inhabits plains, marshes, meadows and sometimes hilly or mountainous country, and prefers to nest near water, high up in trees. The nest is made of branches and lined with earth, paper, moss, sheep's wool and rags, although the bird often takes over the abandoned nests of other species. Eggs, 2, sometimes 3 or 4, are white with reddish flecks. Young remain in the nest for about 45 days. Adults moult during the breeding season. Red Kite (*M. milvus*) has the same distribution as the Black Kite but is considerably rarer.

BLACK KITE

Ornithologists distinguish three groups of kite: the white-tailed kites, swallow-tailed kites and true kites. The Black Kite belongs to the last group and, with the Red Kite, is the only European representative. All are closely related to the Old World vultures (page 139). The Black Kite is dark brown with a shallow forked tail which accounts for its common name in Australia, the Fork-tailed Kite. Predominantly a scavenger, it searches out dead animals or begs food from other raptors such as ospreys, goshawks and peregrines. Even the nest is built or lined with scraps, rags and pieces of rubbish. The Black Kite generally hunts and breeds in groups and often builds its nest near heronries. While the female broods, the male searches and forages for food which the female then distributes among the young.

KESTREL

The Kestrel, a true falcon, exhibits many of the characteristics typical of the group, which also includes sakers, hobbies and merlins. All are highly specialized hunters with long, narrow wings for sustained and swift flight, powerful beaks and feet, and sharp talons. Most have two teeth on the upper mandible with corresponding grooves on the lower mandible. The commonest of the European falcons, the Kestrel is easily distinguished by its high-pitched cry and its habit of hovering in the air while it searches for prey. Unlike other falcons, it seldom captures birds in flight and takes mainly ground-dwelling insects and rodents. The sexes have several important differences in the colour of the plumage. The male has a grey head and rump, and a grey tail with a white band at the tip. The female, illustrated here, has a brown head and a brown tail marked with dark bars.

Order Falconiformes. Family Falconidae. Kestrel (*Falco tinnunculus*), 34cm (13in) long, is found in Europe, Asia and Africa. European species migrate to Africa in winter, as far south as the equator. It inhabits woods, open country and even towns and nests on tall buildings, cliffs, trees and in abandoned nests of other birds such as the Magpie and Buzzard. Eggs, 5–6, yellowish-white with reddish-brown marks, are laid in a shallow scrape. American Sparrowhawk (*F. sparverius*), 22cm (8½in) long, is found in most of the New World. The most wide-spread falcon species is the Peregrine Falcon (*F. peregrinus*) which is found on every continent except Antarctica.

BLACK GROUSE

The Black Grouse is a Eurasian bird of upland moors and meadows. It belongs to the order Galliformes (fowl-like birds) which contains some of the popular game birds – pheasants, grouse and quail. The Black Grouse is polygamous and its courtship is a lively affair. The strikingly coloured cocks gather before the females and perform a noisy dance. The mating call is loud and clear, consisting of a variety of sounds from mews and hisses to gobbles; the accompanying dance is vigorous. The cocks make rapid and jerky movements, jump into the air, stretch out their necks and erect their tails. Competition for the attention of the females often results in vicious fights. Display is followed by mating when the male gathers a harem of three to four hens which remain with him until the end of the breeding season. In common with most other members of the group, the Black Grouse is mainly terrestrial.

Order Galliformes. Family Tetraonidae. Black Grouse (*Lyrurus tetrix*). The cock is 56 cm (22 in) long and the hen is 45 cm (18 in) long. It is found in uplands of Europe and Siberia. The generic name *Lyrurus* refers to the lyre-shaped tail. The male is blue-black with a glossy sheen and spreads his tail feathers during the courting display like the peacock (page 149). The female is brownish-red with brown stripes and a speckled breast. The nest is a depression in the ground and is lined with dried leaves. Eggs, 6–10, are yellow-red with brown spots. The bird eats insects, shoots and berries. New World representatives include the Ruffed Grouse (*Bonasa umbellus*), 45 cm (18 in) long, which is found in woods in northerly eastern and western regions.

145

RING-NECKED PHEASANT

Order Galliformes. Family Phasianidae. Ring-necked Pheasant (*Phasianus colchicus*). The cock may be up to 90 cm (3 ft) long, including the tail which is about as long as the body. The hen is about 60 cm (2 ft) long including the tail of approximately 25 cm (10 in). The female's plumage is of sombre browns and lacks the brilliant colour of the male. Found in central and South-east Asia, Europe, North America including Hawaii, and in New Zealand, it inhabits reed beds and forests in Asia, farmlands edged with trees, thick, low scrub and plains. The nest, made by the female, is usually a scrape on the ground lined with grass and down. Eggs, 6–18, are olive-brown. Young fly at 2 weeks. Diet includes seeds, insects, grains, leaves, roots and worms.

This popular game bird originated in central and South-east Asia and was introduced into Europe some 2000 years ago. Today about 34 varieties are known; these hybrids resulted from breeding the Ring-necked with other races. The plumage of cock pheasants varies according to the strain, but most have a white collar or white markings round the neck. During the breeding season the bird remains hidden in tall grass or other vegetation. The Ring-necked cock establishes a territory of long runs which it travels up and down. Display is accompanied by calling (a version of the *kok kok* cry) and includes erecting the ear tufts, swelling the wattles round the eyes and spreading the wings and tail. When alarmed the pheasant rises almost vertically from the ground, reaching a considerable height, but it is not adept at flying far.

Order Galliformes. Family Phasianidae. Genus *Gallus* includes wild and domestic fowl. Species are polygamous and there are marked differences in plumage of the cocks and hens. All have fleshy crests (combs), ear lobes below the eyes and wattles under the beak. Cocks have spurs on their legs. Red Jungle Fowl (*Gallus gallus*), found in India, Malaysia, Indochina and the Philippines, inhabits mountainous regions in the wild. Some 70 species of domestic fowl are distinguished, divided into 5 groups on the basis of features such as the number of toes (4 in most, 5 in others), plumage and type of comb. Well-known varieties include the American breed, Plymouth Rock and the Leghorns.

DOMESTIC FOWL

Domestic fowl are descended from the Red Jungle Fowl of India and South-east Asia. This bird is known to have been domesticated in India some 5000 years ago and resembles many of the present-day breeds. The cock has a splendid plumage of greens, yellows, reds and browns, a red comb and wattles and long, curved tail feathers. Both wild and domesticated birds have a well-defined social system in which each bird occupies a particular place in the hierarchy or pecking order. Positions are maintained and achieved by chasing and pecking, and among the cocks, fighting, which often becomes a vicious battle. The breeding season lasts throughout much of the year. This explains the success of these birds as egg-producers. Many are also bred by bird fanciers solely for ornamental purposes.

147

GUINEAFOWL

Guineafowl, native to Africa, belong to the family Numididae and are distinguished from other fowl by their spangled plumage and rounded bodies. The Vulturine Guineafowl illustrated here gets its name from the bare blue head and neck which is similar to that of the vultures (page 139). It has a ruff of long black, white and blue striped feathers round the base of the neck and slender, long legs. Other species may have a crest of feathers or a bony casque (helmet) on the top of their head. Guineafowl are gregarious outside the breeding season and assemble in fairly large groups, but during courtship and breeding the male and female isolate themselves. The Common or Helmeted Guineafowl was domesticated by the ancient Egyptians and bred by the ancient Greeks and Romans. It is the ancestor of the present-day domesticated guineafowl raised in many parts of the world.

Order Galliformes. Family Numididae. Wild species are found south of the Sahara and in Madagascar. They inhabit clearings, savannahs and steppes. The nest is a scrape lined with dry vegetation. Eggs, 8–10, yellowish-brown with white spots, are incubated by the female. The diet includes insects and parts of plants such as berries, leaves, grains and tubers. Guineafowl have strong legs and can run at speeds of up to 32km (20 miles) per hour. Vulturine Guineafowl (*Acryllium vulturinum*), 60cm (2ft) long, is found in southern Ethiopia, southern Somalia and north-western Tanzania. Common or Helmeted Guineafowl (*Numida meleagris*) is found south of the Sahara to Cape of Good Hope.

PEAFOWL

A magnificent representative of the pheasant family, the Asian Peafowl is well-known throughout the world, having been imported to most continents as an ornamental bird for parks and gardens. In their native countries peafowl inhabit jungles and forests where they assemble in large groups for most of the year, only pairing off during the mating season. It is during this time that the peacock displays his fantastic plumage most frequently, erecting the fan by means of strong muscles that are located under the rump. These feathers are in fact exaggerated upper tail coverts (feathers that lie over the tail quills) and not the true tail, which is also erectile and is composed of brown feathers. The beauty of the male is also enhanced by the crown of bare-shafted feathers on the top of his head. Display is accompanied by eerie cries, heard most often at daybreak. Terrestrial during the day, most peafowl perch in trees for the night.

Order Galliformes. Family Phasianidae. Two species are usually distinguished: the Common Peafowl (*Pavo cristatus*) of India and Sri Lanka, which has a blue head, neck and breast; and the Javan or Green Peafowl (*P. muticus*) of Indochina and Java, in which the head, neck and breast are green. Both species inhabit dense jungle, usually in hilly regions. The discovery of the Congo Peafowl (*Afropavo congensis*) in Africa, in 1913, established a link between the guineafowl (page 148) and the peafowl. Eggs, 3–10, creamy white, are incubated for 28 days by the female. The young reach adult size in 3 years and may grow for a further 3 years finally reaching a length of 1.6 m (5$\frac{1}{2}$ ft). Peafowl are omnivorous and are known to kill young cobras.

CRANE

Superficially, cranes resemble storks and herons but important differences in anatomy and behaviour place them in the separate order of cranes and rails. Outside the breeding season cranes form flocks which are guarded by a few members who act as sentinels. Breeding pairs keep to themselves, the male assisting the female with incubating the eggs and raising the young. All cranes, including the African Wattled Crane illustrated here, have loud, trumpeting calls which are produced in the long, coiled windpipe. Calls are used to communicate with one another and keep the group together and, in the breeding season, as courting songs between males and females. These birds also perform elaborate dances at all times of the year. They bow stiffly to one another, leap high off the ground and throw sticks into the air which they catch in their bills.

Order Gruiformes: cranes and rails. Family Gruidae, comprising four genera and 13 species. Most make large nests on the ground of dried plant material. The Australian Crane or Brolga (*Grus rubicunda*), lays its two eggs directly on the ground. The young leave the nest and walk and swim on the first day. Diet consists mainly of plants but also of worms, snails, insects and frogs. Wattled Crane (*Bugeranus carunculatus*), 1.5m (5ft) long, is found in eastern and southern Africa, in open, swampy territory. It has two red wattles, one on each side of the throat, and is covered with white feathers. Common Crane (*Grus grus*), 1.2 metres (4ft) long, inhabits northern Europe and Asia and migrates south in winter. It is distinguished by a bald red patch on the crown.

COOT

The family Rallidae is composed of the rails, moorhens and coots – a group of mainly aquatic birds with a wide distribution throughout the world. Although most are water-dwelling species, the toes are not webbed (except in the coots which have partially webbed feet like those of the grebes). Each toe has a separate fleshy lobe which facilitates swimming. The Coot also differs from other members in its gregarious nature, gathering in colonies when it is not breeding. On water the Coot forages for food by diving, staying under long enough to gather aquatic plants and sometimes animals; it also feeds on land. It can be distinguished by its black plumage and white bill and shield above the base of the bill. In flight the narrow white wing bars are apparent. The young have a red bill and orange markings on the head. Their down is black but the early plumage is pale brown above and whitish underneath.

Order Gruiformes. Family Rallidae. Common Coot (*Fulica atra*), 37 cm (15 in) long, is found in Europe, North Africa, Asia and Australia. European birds are mainly resident in their areas but more northerly ones migrate south for winter, travelling at night and often covering 321 km (200 miles) in a night. The nest is made near large bodies of water and is constructed of reeds, sedge and other plants. Eggs, 4–10, are pale brown with reddish and black spots. Common New World species include the American Coot (*F. americana*), 37 cm (15 in) long, which is found from Canada to north-western South America and in the Hawaiian Islands. It inhabits lakes and estuaries. The plumage of both species is black, and they have a white bill and frontal shield.

151

MOORHEN

The Moorhen or Common Gallinule is generally found on small areas of open water surrounded by reeds and rushes. It swims with a characteristic back-and-forth rocking movement and dives below the surface for food. In shallow waters it will squat on the bottom with only the beak showing above the surface. On land it bobs its head up and down and flips its tail as it walks. Curiously, it will also climb the trunks of sloping trees which overhang the water, where it will sometimes make its nest. Unlike its close relative the Coot, but in common with many members of the family, the Moorhen seldom flies. In the past, members of the group led a comparatively isolated life with few, if any, enemies and there was little need for flight. Consequently, as they evolved, many species developed terrestrial habits and a few have lost their ability to fly.

Order Galliformes. Family Rallidae. Moorhen or Common Gallinule (*Gallinula chloropus*), 33 cm (13 in) long, is found on all continents except Australia and Antarctica. It is the most widely distributed member of the family, and inhabits marshes, rivers, ponds and pools, often near human communities. The breeding season may last from spring through summer with one brood in May and a second in late August among birds inhabiting the northern hemisphere. Both parents participate in raising the young, alternately incubating the spring clutch. The second clutch is brooded only by the female, while the male is occupied with caring for the first set of chicks. The nest is located under thick cover or in low branches and is made of stems of local plants.

OYSTERCATCHER

In spite of its name the Oystercatcher rarely feeds on oysters. Its diet consists mainly of limpets, cockles, mussels, worms and insects. The common name in some parts of Europe, the Musselpicker, is therefore somewhat more appropriate. The long, blunt, flattened bill is well-adapted for probing in mud, prying molluscs off rocks and delving into half-open shells. Most Oystercatchers are found along coasts but some now breed further inland. The courtship display by the males is an elaborate affair when two or three birds perform before the females, sometimes on the ground and at other times in the air. At both times the ritual is accompanied by a mating call – a series of *kleeps* followed by a trilling song. Both sexes attend to the preparation of the nest and incubation of the eggs, which lasts for about 28 days. The young leave the nest within two days of hatching.

Order Charadriiformes; waders and gulls. Family Haematopodidae, a single genus comprising six species. They are found on all continents except Antarctica. The European Oystercatcher (*Haematopus ostralegus*), 43cm (17in) long, is found in all parts of Europe. It usually inhabits shingle, sandy beaches, grassy or rocky areas along coasts or, inland, moors and fields. The nest is a scrape, sometimes lined with pebbles or shells and often re-used from year to year by the same pair. Eggs, 2–4, are buff marked with streaks and dots. Black Oystercatcher (*H. bachmani*) is found on the Pacific coast of North America from Alaska to southern California. Sooty Oystercatcher (*H. fuliginosus*) is found in Australia.

LAPWING

Order Charadriiformes. Family Charadriidae. The European Lapwing (Vanellus vanellus), 30 cm (12 in) long is found from Ireland east to the Pacific. It inhabits muddy shores of ponds and lakes, marshy plains, meadows and cultivated land. Diet consists of worms and insect larvae. Its nest is a scrape or depression in the ground. Eggs, 4, brownish with brown to black flecks, are sometimes eaten by other birds (i.e. jackdaws) and by man in countries where they are considered a delicacy. Genus Vanellus is represented in Australia by the Banded Plover (V. tricolor). A close New World relative, also found in Europe, the Ringed Plover (Charadrius hiaticula) is sometimes called the Semi-palmated Plover. It inhabits North America.

In Europe, the name Lapwing usually refers to one species of the plover family which is distinguished from all other members by its long head-crest. It is sometimes also called the Peewit because of its striking *pee-wit* call. From a distance its plumage appears black and white; the breeding plumage of the male is a dark, metallic green on the upper parts, white on the underparts and cheeks, with chestnut undertail coverts. Swift in flight, the males perform aerial acrobatics during the breeding season, somersaulting and plummeting to the accompanying drumming sound made by the wings. Both parents will often feign injury when intruders threaten the eggs or young. The genus *Vanellus*, to which the European Lapwing belongs, includes several species which are found in most parts of the world with the exception of North America, although an occasional European Lapwing wanderer is found even there.

RUFF

Although the legs and bill are much shorter, in its winter plumage the Ruff resembles its close relative the Redshank (page 157). During the breeding season it is completely distinguishable from the Redshank and all other birds by the long erectile feathers round its neck which compose the ruff, or frill, and by the feathery tufts behind the ears. The female, or reeve, is smaller than the male, lacks the frill and is greyer in colour. Courtship is an elaborate affair. The males gather before dawn on special areas called leks. Each one takes up a special area and defends it from other males by threatening postures and mock battles. With the appearance of the females, the Ruffs display their frill and lower their heads. Pairing is initiated by the females who choose their own mate. The actual mating takes place either on the hill or at some distance in more isolated surroundings. The nesting site is usually away from the male's territory.

Order Charadriiformes. Family Scolopacidae, woodcocks, sandpipers and curlews. The male Ruff (*Philomachus pugnax*) is 28 cm (11 in) long; the female, or reeve, is 23 cm (9 in) long. The Ruff is found in northern parts of Europe and Asia and spends winter in Africa and southern Asia. Occasionally wanderers are found on the east coast of North America. It inhabits salt marshes, estuaries and wet marshes near coastal shores. The nest is a shallow depression in grassy tussocks. Eggs, 3–4, are grey to olive-brown with grey or dark spots. Colour of the male's breeding plumage varies from one individual to another. The ruff may be various browns, black or white, sometimes with bars or other markings. Numbers of Ruffs in England have declined.

COMMON SANDPIPER

The sandpipers exhibit incredible stamina during flight and some species migrate a distance of over 9600 kilometres (6000 miles). The Common Sandpiper, like all species, breeds in uplands in the high latitudes of the northern hemisphere. It migrates for the European winter to the subtropics and tropics as far south as Australia and Argentina. Breeding takes place in early summer when the male performs his courtship display in the air. He is sometimes accompanied in the nuptial flight by the female who flies with him willingly; or the male may pursue the female in the air or on the ground. In defence of the nesting site and young, both parents may show aggressive behaviour by bold attack or they may attempt to divert intruders by feigning injury.

Order Charadriiformes. Family Scolopacidae. Common Sandpiper (*Tringa hypoleucos*), 20 cm (8 in) long, is found in great numbers in most parts of northern Europe, and is rarer in the Netherlands, Denmark and Belgium. It inhabits upland wood-fringed lakes, streams and rivers or sheltered islets near the coast. In winter it migrates south to South Africa, Australia and Argentina. The nest is a depression hidden among vegetation. Eggs 3–5, are yellowish to grey with dark spots or flecks. The diet consists of insects (often caught in flight), worms, snails and crustaceans. Similar New World species, the Spotted Sandpiper (*T. macularia*), 20 cm (8 in) long, breeds in most parts of North America, and migrates in winter to the tropics.

REDSHANK

Order Charadriiformes. Family Scolopacidae. The Redshank (*Tringa totanus*), 28cm (11in) long, is found in Eurasia from the British Isles as far east as China. It migrates south in the autumn, sometimes as far as South Africa. It inhabits sand dunes, salt marshes, shingle shores, wet grassland and heath in summer, estuaries and mud banks in winter. The nest is a depression lined with dried grasses and usually has a protective canopy made of interwoven or bent stems of surrounding plants but on open shores the nest may be exposed. Eggs 3–5, are pale brown with grey marks and dark brown spots. Both parents incubate the eggs and in the early stages care for the young; later they are protected only by the male.

A long, red-legged bird of the genus *Tringa*, the Redshank is a common wader in temperate regions of Europe and Asia. Found along coastal shores or on wet, inland grasslands and heaths, its presence can often first be detected by its piercing alarm call. Both male and female have a varied calling repertoire; the male sings during display flights, ascending rapidly into the air and then gliding down towards the ground. In flight the distinguishing white bars on the wings and the grey barred tail are visible. The slightly larger Spotted Redshank is similar in habit and appearance, differing mainly in the colour of the plumage which is black with white spots in summer and greyish in winter; it lacks the wing bars and has a longer bill. The Spotted Redshank is rarer than its relative and has a more restricted distribution. Both birds are migratory although they do not cover the distances of some other members of the family.

157

Order Charadriiformes. Family Glareolidae, pratincoles and coursers. Common Pratincole (*Glareola pratincola*), 25cm (10in) long, is found in southern Europe, Africa, and southern Asia. It inhabits marshy areas and dunes near coastal waters and nests in small colonies. Eggs, 2–3, are earth-brown with dark markings. The nest is a shallow depression made on dry ground and its diet consists of insects, particularly dragon-flies and grasshoppers. (In South Africa they are called Locust Bird.) Coursers are also insectivorous; they have shorter wings and do not migrate any great distance, except for the Egyptian Plover (*Pluvianus aegyptius*). This courser inhabits sandy river banks.

COMMON OR COLLARED PRATINCOLE

The Common or Collared Pratincole is found along coastal shores in warm temperate regions of the Old World and is classed as a wading bird (the front toes are webbed) although it spends much of its time on the wing. The air is its natural element and it can often be seen circling, seemingly endlessly, in the sky or skimming low over the ground with gaping beak, hawking insects from the air. It also runs on the ground and picks up food from the soil. The Common Pratincole has a sharply forked tail, with tail streamers as much as 5 centimetres (2 inches) longer than the inner feathers, a characteristic that has given it the common name Sea-swallow. The wings are long and pointed like those of the tern. The Pratincole often migrates long distances in flocks of up to 1000 individuals; some birds of southern Asia make an annual journey as far as Australia.

BLACK TERN

Order Charadriiformes. Family Laridae, gulls and terns. Black Tern (*Chlidonias niger*), 24cm (9½in) long, is found in Europe, Asia Minor and central Asia in the Old World, and on the coasts and inland lakes of North America in the New World. It migrates to subtropical and tropical waters. It breeds in small colonies or in isolated pairs. The nest is built on water, made of a floating mass of vegetation, or on tussocks. Eggs, 2–3, are yellowish, brownish or olive green with greyish and dark brown flecks. Arctic Tern (*Sterna paradisaea*) is 38cm (15in) long. Its plumage is white with a grey mantle, black cap and blood-red bill and it is found on coasts of northern Europe, Asia and North America.

Terns are graceful birds that live mainly along coastal shores, although some species breed further inland near freshwater. Masterful fliers with long, pointed wings, they glide and swoop on the wind, seldom alighting on water for any length of time. They walk laboriously on land. The bill is pointed and refined, the feet webbed and legs short, and the tail slightly forked. The Black Tern of Eurasia and North America is almost entirely black and dark grey in the breeding season. At other times of the year it takes on white markings on the forehead, neck and underparts. This tern feeds by skimming low over the water and gathering its food from the surface. At the close of the breeding season the Black Tern migrates to tropical waters. The Arctic Tern, a close relative, makes one of the longest migrations of any bird from Point Barrow near the North Pole to the Ross Sea in Antarctica.

COMMON TERN

The striking appearance of the Common Tern is remarkably similar to that of the Arctic Tern (page 159) and these two species are often confused. Both have bright red bills but in the Common Tern the bill is tipped with black and the legs become reddish-brown in winter rather than black. In the breeding season it inhabits much of the northern hemisphere and most are found inland. However, in Scandinavian countries and Britain it is more frequently seen along sea coasts. Courting takes place on the ground or in the air where one bird will fly carrying a fish in its beak. It is usually joined by another and the two wheel and glide, transferring the fish, when they land, to the other birds in the colony. Both parents incubate the eggs and care for the young which are fledged at about four weeks. The Common Tern is carnivorous and feeds on animals that float just below the surface of the water.

Order Charadriiformes. Family Laridae. Common Tern (*Sterna hirundo*), 34cm (13¼in) long, is found in Europe, Asia and North America. It migrates to sub-tropical and tropical waters in South Africa, north-western Australia and South America. It inhabits sandy beaches, rocky islands, lakes and moors. The diet consists of small fish, worms, crustaceans and molluscs. Insects are also taken on land. When food is scarce, the Common Tern and most other species will steal fish from other birds. It nests in large colonies (terneries), making a shallow scrape, usually unlined. Eggs, 2–4, are creamy-white to greenish-brown, and marked with dark blotches.

GREATER AND LESSER BLACK-BACKED GULLS

Noisy and gregarious, gulls gather in large groups on sea and ocean coasts in every part of the world including the polar regions. However, the common name seagull gives a false impression of their habits, as many are also found inland and none of the gulls ever ventures too far from land. The Lesser and Greater Black-backed Gulls of the northern hemisphere exhibit many of the characteristics of the family: webbed feet, a hooked bill and long, flexible wings which make flying and gliding an effortless activity. The Greater Black-backed Gull is the largest species found in northern Europe and is very similar to the smaller Lesser Black-backed Gull. Both have blue-grey or black upper parts with a pure white head and underparts.

Like all gulls they are scavengers, frequenting harbours, ports and rubbish dumps to forage for food.

Order Charadriiformes. Family Laridae. Greater Black-backed Gull (*Larus marinus*), 66 cm (26 in) long, with a wing span 1.5 metres (5 ft) wide, is found in the North Atlantic, as far south as Brittany in the Old World, and on the east coast of North America. It inhabits rocky coasts, offshore islands, inland lakes and moors. The nest is made of seaweed, dried grass and sticks. Eggs, 2–4, are greyish to brownish with brown flecks. Lesser Black-backed Gull (*L. fuscus*), 53 cm (21 in) long, is found from Iceland east to the White Sea (including Great Britain, France and Scandinavia). It inhabits coastal cliffs and islands, inland rivers, lakes and moors. The nest and eggs are similar to those of the Greater Black-backed Gull's. Both species prey on other birds and their eggs.

BLACK-HEADED GULL

The Black-headed Gull is the most numerous and best-known of the European gulls. It was once restricted to the Old World but today it is not uncommon for some to stray to the east coast of North America during the winter months. In their native lands of Europe and Asia they range from coastal and inland waters to large city reservoirs. Red-billed and red-footed, the Black-headed Gull has black wing tips edged with a narrow white band. During the breeding season the head has a dark brown hood; in winter it is white with a small, dark spot on either side. As with other species, pairing is preceded by some fighting between the breeding males. Nests are most frequently made in large colonies, and the whole breeding area is filled with the gulls' noisy cries most of the day. Grebes, terns and other birds may also nest in the same vicinity. Both parents share in the incubation of the eggs.

Order Charadriiformes. Family Laridae. The Black-headed Gull (*Larus ridibundus*), 38 cm (15 in) long, is found in Europe and in temperate parts of Asia and migrates south to North Africa, southern Europe and southern Asia. It inhabits sand dunes, salt marshes, rivers, lakes and moorland waters. Some vegetable food is taken but the main diet consists of fish, worms, eggs of waterfowl, insects and offal. The nest is made on wet or damp ground or on water, and is built of stems of aquatic plants; it occasionally consists of a mere scrape. Eggs, 2–3, are pale blue or green to dark brown with dark brown blotches. Young have reddish-brown and dark brown upper parts with pale underparts, a dark tip on the flesh-coloured beak and two black patches on the throat.

HERRING GULL

Order Charadriiformes. Family Laridae. Herring Gull (*Larus argentatus*), 56 cm (22 in) long, is found in Iceland, the Faeroe Islands, Great Britain and Ireland, Scandinavia and the Baltic, and on the east and west coasts of North America. It inhabits sand dunes, rocks and cliffs along shores, harbours and ports. The nest is sometimes built on buildings and infrequently in trees. It may be a scrape in the ground or a large structure made of seaweed, moss and grass. Eggs, 2–3, are olive-green to brown with black flecks. Both male and female are fierce protectors of their eggs and young. The diet consists of offal, small mammals and birds, eggs, earthworms, insects, grains and roots, fish and crustaceans.

The Herring Gull is a common bird of the Old and New Worlds and is one of the most interesting of the gulls. It is sometimes confused with the Common Gull, especially when young. Its pale grey wings are tipped with black, and the yellow bill has a red spot on the lower mandible; the legs are pink. A scavenger, gleaner and hunter, the Herring Gull steals from other birds, taking their food, eggs and even killing the young, including the young of its own species. It forages in fields for grains and roots and feeds on waste from fishing boats and rubbish dumps. It also dives into the water to catch fish and other aquatic animals, dropping shellfish from a great height to break open the shells. Unlike other gulls, the Herring Gull sometimes pairs with the mate from previous years. The male takes a more active part in brooding and defending its young. The female usually makes the first advances during the pairing.

PUFFINS

Puffins belong to the family Alcidae, which includes the auks and guillemots. These birds fulfil the same role in the northern hemisphere as the penguins do in the southern hemisphere and in many respects the auks are similar to the penguins, although they are unrelated. Auks are not flightless but most species are more at home in the water than in the air. The wings, which are short, propel the body through the water and the webbed feet are used for steering. Like penguins, the auks and their relatives walk upright on land, and the same black-and-white colouring predominates in most species. The puffin is easily distinguished from all other members by the parrot-like bill which has earned it the common name Sea-parrot; in the breeding season the bill is decorated with bright red, yellow and black plates. The plates are shed in winter (leaving the bill smaller and duller) as are the blue-grey plates above and below the red-rimmed eyes. The legs are orange-red in summer and yellower in winter. Both male and female are alike although the male is usually heavier.

The Common Puffin, found in the North Atlantic regions of Iceland, Greenland and the Arctic archipelagos, is a strong flier and spends the winter at sea. Early spring marks the beginning of the breeding season when the Puffins arrive in enormous numbers at the nesting sites. The rookeries are burrows made in grassy slopes and cliffs, with a nest at the end; some tunnels may be as long as 2 metres ($6\frac{1}{2}$ ft) and all of them are placed extremely close together. Holes among rocks and cracks in cliff faces are also used as nesting sites. Upon arrival both males and females take part in a courtship display, pairing already having taken place at sea. This consists of dances on water and land, head-bobbing, bill-clashing and rubbing. One egg only is laid at the bottom of the burrow. Both parents take turns at incubating for some 42 days. The parents gather food for themselves and their young by flying some distance out to sea. The chick is fed fish, mussels and sea-urchins almost constantly, its weight increasing rapidly until at six weeks it is heavier than its parents. The juvenile is then left alone and lives off its food reserves for several days while learning to fly and hunt for itself at sea. Many puffin chicks are taken by the Greater Black-backed Gull.

Order Charadriiformes. Family Alcidae, auks, guillemots and puffins. Common Puffin (*Fratercula arctica*), 30 cm ($12\frac{1}{2}$ in) long, is found in north-western Europe (as far south as Brittany), Iceland, Greenland, the Arctic archipelagos, Faeroe Islands and in north-eastern North America as far south as New England. It inhabits mainland and island cliffs and slopes. Nests are sometimes made in abandoned rabbit holes. The single egg is white with pale grey and brown spots. Diet is fish, especially sand-eels, and is gathered by diving. Characteristically the Puffin holds up to 11 fish in its beak at one time. A similar species is the Tufted Puffin (*Lunda cirrhata*), 30 cm ($12\frac{1}{2}$ in) long, found in northern Pacific regions off Asia and North America, southwards to California.

Order Columbiformes, pigeons and sandgrouse. Family Columbidae, comprising over 280 species. The names pigeon and dove are interchangeable. Collared Dove (*Streptopelia decaocto*), 30 cm (12½ in) long, is found throughout most parts of Europe. It inhabits rural and urban areas near human communities and is resident wherever it breeds. The nest, made of twigs, is usually in trees at least 1.8 metres (6ft) above ground or in bushes but always well-hidden. Eggs, 2, are glossy white. There may be up to 5 broods in one year. Both male and female incubate the eggs; the males at night, and the females during the day. Young leave the nest when about 15 days old. Diet consists of seeds, grains, fruit and berries.

COLLARED DOVE

The Collared Dove – sometimes called the Collared Turtle Dove – belongs to the family of pigeons, Columbidae. Originally found only in Eastern Europe, during the past 40 years it has spread to most parts of Europe including the British Isles. In common with all pigeons it has a swollen membrane (cere) at the base of the short beak, which extends over the nostrils. The plumage is almost identical in both sexes: grey-brown over most of the body, with black and white undertail and narrow half-collar round the back of the neck. The male is usually slightly larger and darker than the female. The Collared Dove generally lives near human communities, in parks and gardens in cities, and in villages and farms in rural districts, where it often feeds with domestic fowl. The cooing call is accompanied in the breeding season by a bowing display which takes place on a perch among several individuals.

TURTLE DOVE

The Turtle Dove frequents hedges and thickets bordering fields and meadows, parks and open woods. The smallest European member of the pigeon family, it has a small head and long neck and is altogether a more graceful bird than most other members of the family. The plumage is brownish red with variegated upper parts and several black and white bands on the back of the neck; the fan-shaped tail has a white margin at the tip. During the breeding season, usually from May to June, several birds gather to form small colonies. Display includes the characteristic bowing between pairs, which is performed with the crop inflated and the bill pointing downwards, and flights by the male in which he rises steeply into the air from his perch, circles and returns to his original position. The Turtle Dove has a soft, purring call and the low *trurr trurr* can be heard throughout the summer wherever it breeds.

Order Columbiformes. Family Columbidae. The Turtle Dove (*Streptopelia turtur*), 28 cm (11 in) long, found in southern parts of Europe, western Asia and North Africa, inhabits bushy and open wooded country and farmland. It nests in small colonies in hedges, thickets and young coniferous plantations but not in extensive woods. The nest is usually built quite close to the ground and is made of twigs. Eggs, 2–3, glossy white are incubated by both male and female. There may be two broods a year. Young are fed on 'pigeon milk', a curd produced in the lining of the parents' crop. They leave the nest after approximately 15 days. Diet consists of seeds and other plant parts but some molluscs are also taken.

STOCK DOVE

Sometimes called the Blue Rock, the Stock Dove is often confused with the Rock Dove, although the latter is found only along coastal shores while the Stock Dove inhabits woods, forests and occasionally sand dunes and sea cliffs. It has bluish-grey upper parts with metallic green patches on either side of the head, a pinkish breast and two faint narrow black bands on the wings; the tip of the tail also has a faint black band. The Rock Dove is distinguished by broad, conspicous black bands on the wings and, in flight, by the white rump. The courtship display is similar to that of the Collared and Turtle Doves (pages 166, 167) but Stock Doves also take part in peculiar flights when they slap each other with their wings. These birds are swift in flight. Most do not travel very far from their breeding sites, although the more northerly doves move further south for the winter.

Order Columbiformes. Family Columbidae. Stock Dove (*Columba oenas*), 33 cm (13 in) long, is found in Europe, Asia and north-west Africa. It inhabits woods, forests, and sometimes sand dunes and sea cliffs. Nests are made in tree holes, especially those made by woodpeckers, in nest boxes and occasionally in rabbit holes or in mud or sand banks. The hole may be lined with a few twigs and leaves. Eggs, 2–3, are glossy white. There are two broods, and sometimes three or four a year. Rock Dove (*Columba livia*), 33 cm (13 in) long, is found in southern Europe, North Africa, central and southern Asia and North America. The ancestor of the Domestic Pigeon, it can fly at speeds of up to 185 km (115 miles) per hour.

BURROWING PARROT

Order Psittaciformes, parrots and parakeets, comprising only one family, the Psittacidae with around 325 species. The tribe Psittacini contains the true parrots. Burrowing Parrot (*Cyanoliseus patagonus*), 45cm (18in) long, is found in South America, in Uruguay, Chile and parts of Argentina, and inhabits open plains. The nest is made in the ground at the end of a tunnel which the bird excavates itself. Eggs, 4, brilliant white, are incubated by the female for about 26 days. It migrates north for the winter.

Although many parrots conform to the popular conception that they live in trees in tropical jungles, some, such as the Burrowing Parrot, inhabit relatively treeless areas. Others even live high in the mountains where the climate is quite harsh during the winter. The Burrowing Parrot is a bird of the South American plains and is found in Uruguay, Chile and parts of Argentina. The plumage is dark olive-brown on the upper parts, neck and head; the breast, upper tail coverts and rump are bright yellow. It flies in large flocks searching for food and feeding on parts of plants such as seeds, roots and buds. Unlike most parrots, which nest in tree holes, the Burrowing Parrot nests in the ground. The breeding colonies often cover a fairly wide area. Long tunnels are dug in sand or mud banks and a chamber is enlarged at the end, where the female lays her eggs.

GOLD AND BLUE MACAW

Parrots are found in warm temperate and tropical regions throughout most parts of the world, Europe being the only continent without native species. The greatest concentrations of these birds occur in Austral-asia and in South America, the home of the macaws. The Order Psittaciformes (parrots and parakeets) is a large group of mainly arboreal birds whose feet and beak are adapted for both climbing and feeding. In common with other climbing birds such as the woodpeckers (page 180) and the nuthatches (page 206) the toes are zygodactylous, i.e. the second and third toes point forwards and the first and fourth toes point backwards. In parrots, the toes are flexible enough to be used as hands which can grasp food as well as tree trunks and branches. The beak is distinctive: both the upper and lower mandibles are curved, with the lower mandible fitting into the upper one. In addition the upper mandible is jointed at the base and can be moved up and down. The true parrots include some of the best-known birds of this order: the macaws, lovebirds, parakeets, budgerigars and cockatoos. Many of these are excellent mimics and have highly developed memories, their ability to 'talk' also being partly due to the fleshy tongue. The African Grey Parrot is particularly favoured as a candidate for speech training, while the Australian budgerigars are bred and sold as popular cage birds, both for their bright colours and their ability to mimic human speech.

The Gold and Blue Macaw is found in forests and woods from Panama to Argentina. It is typical of the other 14 species of the genus *Ara*, all large, brightly-coloured birds with long pointed tails, large hooked beaks and a harsh screaming call. Ridges beneath the tip of the upper mandible are used to file notches in hard shells which the Macaw then cracks with its strong beak. This bird gathers in large flocks to feed, taking fruit and berries as well as nuts; and breeds in colonies, making its nest in hollow trees. Pairs are monogamous and probably mate for life, which is unusually long, most parrots being renowned for longevity.

Order Psittaciformes. Family Psittacidae. The Gold and Blue Macaw (*Ara ararauna*), 90 cm (3 ft) long, is found in South America, from Panama to Argentina and inhabits tropical rain-forests. Diet of fruit, berries and seeds. It is one of the most commonly-seen birds in zoos. African Grey Parrot (*Psittacus erithacus*) is a popular cage parrot. Grey with a bright red tail, it is found in western and central Africa. Budgerigar, (*Melopsittacus undulatus*), 18 cm (7 in) long, is found in Australia. In the wild it has green underparts and tail; the upperparts are yellow with black stripes. Domestic breeds exhibit a wide variety of colours including pale blue.

170

CUCKOO

All 130 species of cuckoo are arboreal but they perch rather than climb. Curiously however, the arrangement of the toes is like that of climbing birds, with two toes in front and two behind (see pages 170 and 180). The Common Cuckoo of Europe and northern Asia is a summer visitor that arrives in early spring to breed. In common with many other members of the family it is a social parasite that lays its eggs in the nests of small birds. The Meadow Pipit is one of the most usual hosts but others include the Willow Warbler, Robin and Wren (see pages 199, 194 and 186 respectively). The birds chosen are all either insect-eaters or birds that feed their young on insects – the necessary diet for young cuckoos. Following mating, the hen lays a single egg in the chosen nest, having previously disposed of one of the eggs laid by the host bird. This procedure is repeated at intervals of two days until the hen has laid 12 eggs, each in a different nest.

Order Cuculiformes, cuckoos and turacos. Family Cuculidae. Common Cuckoo (*Cuculus canorus*), 30cm (12in) long, is found in Africa and migrates north in the spring to Europe and northern Asia. It inhabits woods and open ground. Diet consists of insects and their larvae, especially caterpillars. The nestling is born blind and naked. Soon after hatching, it ejects the eggs or young of the host bird from the nest so that it is the sole inhabitant. It is fed by the foster parents for about 3 weeks in the nest and a further 3 weeks after it has fledged. The mating call of the male is the familiar *cuck-oo* and the female responds with a chuckling song. The Yellow-billed Cuckoo (*Coccyzus americanus*) and Black-billed Cuckoo (*C. erythropthalmus*) are not parasitic.

LONG-EARED OWL

Owls are divided into two families: the barn owls, Tytonidae, consisting of ten species, and the typical owls, Strigidae, of which there are 120 species distributed throughout the world. The Long-eared Owl of the Old and New Worlds is a typical owl that inhabits woods, parks and forests, particularly coniferous forests. Its mottled plumage, usually chestnut and brown, resembles the bark of trees and effectively camouflages the bird when roosting during the day; its presence, however, can often be detected by the circle of pellets and droppings on the ground below the roost. The two feathery tufts on the head (from which the Long-eared Owl gets its name, although they are not really ears) are erectile and lie flat on the head during flight. Long-eared Owls infrequently build their own nests, preferring to take over the abandoned nests of birds such as the Magpie. In most years only one clutch is laid.

Order Strigiformes, owls. Family Strigidae. Long-eared Owl (*Asio otus*), 34cm (13½in) long, found in Europe and northern Asia, North America and North Africa, inhabits woods, parks, and forests. It nests rarely on the ground, usually in trees, on cliffs or buildings. Eggs, 4–6, white, are incubated by the female. The young remain in the nest for about 24 days. Diet consists of small rodents, hunted at night and eaten whole. When prey is scarce it will also take small birds. The call is more a mournful *coo* than a hoot. It barks during the breeding season and displays by flying through the woods and clapping its wings.

LITTLE OWL

The typical owls, Strigidae, are skilful, silent hunters perfectly adapted for searching out and capturing prey. Soft plumage and rounded wings enable them to fly noiselessly through the night although some, such as the Little Owl, have diurnal habits and hunt during the day or at dusk. In the Strigidae the senses of smell and sight are highly developed. The large eyes, set in the front of the head, restrict the birds' visual range but this is compensated for by the ability to turn the head round between 180 and 270 degrees. Owls are long-sighted and have difficulty focusing on objects immediately in front of them but they can feel by means of sense bristles situated around the beak. The typical owls are distinguished from barn owls by the disc-shaped facial mask and by shorter, partially feathered legs. (Barn owls have a heart-shaped mask and their long legs are completely covered with feathers.) Typical owls are equipped with strong feet, sharp claws and beak, and feed by grasping prey in the claws and either swallowing it whole or tearing the flesh into pieces, depending on the size of the prey.

The Little Owl is native to southern Europe, North Africa and Asia but has been introduced into more northerly European countries including Great Britain, where it now breeds successfully and is one of the most commonly seen species. In other countries it has not fared as well, owing mainly to the cold winters. It is small and plump: other distinguishing features include the brown and white mottled plumage, short tail and wide, flattened head. The facial mask is more indistinct than in other species. The Little Owl usually lives near human communities, especially farms. Although most of its hunting is done at dawn or dusk it may also be seen during the day, flying low over the ground or perching on a pole or post. It has a peculiar habit of bobbing up and down when disturbed or alarmed and it yelps when it is excited. The usual call is a *kiew kiew*. Unlike the Long-eared Owl (page 173) it has a varied diet and hunts small mammals, rodents, birds and insects. The male's display flight takes place over the nesting site which is usually in tree holes or in walls and lofts of old buildings. The young have short, white down and are fed by both parents for the first four weeks.

Order Strigiformes. Family Strigidae. Little Owl (*Athene noctua*), 22cm (8½in) long, is native to southern Europe, Asia and North Africa, and it inhabits open country with trees. The nest is made in holes, in trees and buildings, and sometimes on the ground or in nest boxes. No nesting materials are used. Eggs, 3–6, are matt white. Diet consists of small mammals, rodents, birds, insects and occasionally earthworms. It hunts by pouncing from its perch and hovering before capturing its prey. Typical owls of North America include the Great Horned Owl (*Bubo virginianus*), 62cm (24½in) long, found from Alaska and northern Canada to South America. Its habitats vary from rocky country to parks and gardens. The family Tytonidae includes the Barn Owl (*Tyto alba*), 35cm (14in) long, which is found in Europe, Asia, Africa, the United States, South America and Australia.

KINGFISHER

The Order Coraciiformes consists of around 190 species of widely distributed birds which, besides the kingfishers, includes the hoopoes and hornbills (pages 178, 179). All are long-billed, short-legged birds with bright, conspicuous plumage that nest in burrows. In most, two or three of the front toes are joined along part of their length. The family of kingfishers, distributed throughout most of the world, is divided into three subfamilies. The true kingfishers include the Common Kingfisher (illustrated here) and the Kookaburra of Australia. The Common Kingfisher is a solitary bird that is generally found along slow-moving rivers and streams with steep banks suitable for excavating the burrow. It is known for its extremely rapid flight, during which it utters the shrill *chee chee chee* cry. It generally hunts by perching on an overhanging branch and watching the water for the movement of small fish. When a fish is sighted the Kingfisher dives down into the water, grasps the fish in its bill and takes it back to the branch, usually swallowing it head first. At other times it will hover for a short period over the water before it plunges beneath the surface. The nesting burrow is dug with the bill to a length of about 90 centimetres (3 feet) and inclines upwards. At the end a nest chamber is made; it is unlined but usually scattered with fish bones. While the female incubates the eggs, the male brings in food and removes the droppings. The young remain in the burrow for about three and a half weeks. After they have left, a second clutch is incubated, often in the same burrow.

The Kookaburra is the largest member of the kingfisher family and one of the best-known Australian birds. Sometimes found scavenging in towns, it also preys on small mammals, young birds, large insects and snakes. It makes its nest in hollow trees or in termites' nests. The loud, maniacal, laughing cry is most often heard in the early morning. In the New World the kingfishers are represented by the Cerylinae. The most common is the Belted Kingfisher which has blue-grey upper parts, white underparts and a blue-grey band across the breast. It is found in both eastern and western regions of the United States and Canada and inhabits inland waterways.

Order Coraciiformes. Family Alcedinidae, true kingfishers. Common Kingfisher (*Alcedo atthis*), 15 cm (6 in) long, found in southern Europe, Asia, North Africa, inhabits rivers and streams. The nest burrow is dug in sand or muddy banks. Eggs, 6–10, white, are laid in the nest chamber. Two, sometimes three, broods are laid each year. The young hatch naked, are reared by both parents and leave the burrow at about 4 weeks. Diet consists of small fish, with some water beetles and dragonfly larvae. Kookaburra or Laughing Jackass (*Dacelo gigas*, sometimes classed as *D. novaeguineae*), 42 cm (17 in) long, is found only in Australia, and inhabits towns and open country. The plumage is dark brown on upper parts, white underneath, with a white band between head and body, and a dark stripe through the eyes. The diet consists of snakes, insects, small mammals and young birds. Belted Kingfisher (*Ceryle alcyon*), 32 cm (12½ in) long, is found in North America and eats fish.

HOOPOE

The Hoopoe is a conspicuous bird easily distinguished from all other species by its bright, orange-brown plumage and the black-tipped crest on its head. In flight the rounded black and white barred wings are particularly noticeable. Solitary and timid, the Hoopoe alternately fans out and contracts the crest when it is excited. The long bill is used to probe in soil, mud and among refuse for insects, worms and small lizards. Both sexes are alike and the young resemble their parents. Native to Africa, Madagascar and southern regions of Eurasia, the Hoopoe is an occasional summer and autumn visitor to Great Britain, where it is sometimes seen on southern coasts. It generally prefers open country and breeds in well-wooded areas in old trees or in buildings. Nesting sites include the hollows of trees, roofs and rock piles. The nest is very foul and exudes a strong odour.

Order Coraciiformes. Family Upupidae, comprising only one species, the Hoopoe (*Upupa epops*). Its length is 28 cm (11 in) including the bill which is about 6 cm (2½ in) long. It is found in Africa, Madagascar, southern Europe and Asia and inhabits open country with some cover. The breeding call is *hoop-poo-poo*. It nests in holes in trees, either natural or made by the woodpecker, in holes in the ground, buildings and in rock piles. No nesting material is used. Eggs, 4–9, at first white or greenish-grey, are later stained light brown from droppings. The young leave the nest at about 26 days. It feeds on worms, insects, and other small animals including lizards.

HORNBILL

Order Coraciiformes. Family
Bucerotidae, consisting of 45
species. Hornbills are found in
tropical regions of Africa and
Asia, sometimes near human
communities. They inhabit well-
wooded areas but some are
ground-dwellers. Diet includes
fruit, seeds and insects. Hornbills are monogamous and
mate for life. Asian species include the Great Hornbill (*Bu-
ceros bicornis*), 1.3 metres
(52 in) long, which is found in
India, Malaysia and Indonesia.
It has black and white plumage
and a yellow bill. The Red-billed
Hornbill (*Tockus erythrorhyn-
chus*), 50 cm (20 in) long, is an
African bird found on the open
plains. Plumage is black with
white markings on the upper
parts, and white below. The bill
is uniformly red or black at the
base.

Some 45 species of hornbill inhabit the tropical regions
of Africa and Asia. Their most conspicuous feature is
the downward curving beak which in most species has a
horny casque or helmet protruding at the top. The bill is
either hollow or filled with spongy bone tissue. Most
hornbills nest in tree cavities and their breeding behaviour is unique to the family. After mating, the female
enters the cavity where she will lay her eggs. The hole is
then filled in by both male and female with clay, mud
and sometimes regurgitated food, leaving only a small
opening through which the female can feed and pass
droppings. The male gathers food for his mate and the
young, which take several months to reach maturity.
During the brooding period the female undergoes a
moult. She emerges when the young are half-grown and
assists the male in feeding the chicks. This behaviour
probably safeguards the eggs and young from predators.

GREAT AND LESSER SPOTTED WOODPECKERS

Order Piciformes. Family Picidae. Great Spotted Woodpecker (*Dendrocopos major*), 23 cm (9 in) long, found in Europe and Asia, inhabits parks, large gardens and woods. It is usually resident where it breeds, but some northerly birds move south for the winter. It nests in coniferous and broadleaved trees. The nesting hole is usually made at least 4 metres (13 ft) from the ground. Eggs, 4–8, are glossy white. Young hatch naked after 12 days incubation and leave the nest after 3 weeks. The diet consists of insects, seeds and sap from lime and other trees. Lesser Spotted Woodpecker (*D. minor*), 14 cm (5½ in) long, found in Europe and Asia, occurs in similar habitats to those of the Great Spotted Woodpecker but avoids coniferous woods.

The true woodpeckers are arboreal climbing birds of the Old and New Worlds. The Great and the Lesser Spotted Woodpecker are typical representatives of the family. Both have a large head in relation to the body, short legs and sharply clawed feet; the first and third toes point forwards and the second and fourth toes point backwards. The strong, horny beak has a chisel-shaped tip. They feed by clinging to the tree trunk, supported by their stiff tails, hammering at the bark and gathering insect larvae and other animals with the long, sticky, hook-tipped tongue. The Great Spotted Woodpecker (illustrated on the right) is the larger of the two birds and can be distinguished by the large white patch on either wing and the crimson under the tail. The female lacks the crimson marking on the head. The Lesser Spotted Woodpecker has black and white bars on the wings.

SKYLARK

Order Passeriformes, comprising some 5000 species of perching birds. Family Alaudidae includes 75 species of larks. Skylark (*Alauda arvensis*), 18 cm (7 in) long, is native to Europe, temperate Asia and north-west Africa. It has been introduced into Australia, New Zealand and Vancouver Island in British Columbia. It is found on large areas of open ground such as moors, sand dunes and heaths. The nest of grass is built in a depression on the ground and is hidden by low-growing vegetation. Eggs, 3–5, are grey, yellow or red with brown spots. There are usually two broods a year and sometimes three. The Cuckoo (page 172) will sometimes lay its egg in the Skylark's nest. Diet consists of seeds and insects. The Horned Lark is native to North America.

The Skylark is best known for its soaring song flight which the male performs above the nesting site during the breeding season. Rising almost vertically, it hovers high in the air singing its melodious warbling song, drops suddenly with wings spread and then rises again. This ritual is repeated several times. The family of larks, the Alaudidae, are ground-nesters whose plumage is inconspicuously earth-coloured. The Skylark has brown upper parts streaked with darker browns, white outer tail feathers and white underparts. Like many other larks, it has a crest but this generally lies back against the head. Other family characteristics include a long, clawed hind toe (all perching birds have three toes in front and one behind), broad, long wings and a conical-shaped beak. Skylarks gather in large flocks in the winter but pairs keep more to themselves in the breeding season, and fights between males are not uncommon.

SWALLOW

The family Hirundinidae contains 75 species of swallows and martins; all are slender birds with elongated bodies, long pointed wings and forked tails, characteristics which make them swift and graceful fliers that can soar high above the ground. They are insectivores that feed on the wing. Keen sight enables them to spot minute insects, which they gather through their wide gape. They are truly aerial birds. Most are usually seen on the ground only at the beginning of the breeding season when they collect building materials for the nest; movement over the ground is generally laborious and clumsy. Swallows and martins are found in most parts of the world with the exceptions of the polar regions and New Zealand.

The Common Swallow breeds in Europe, Asia and in North America where it is known as the Barn Swallow. It is a beautiful little bird with glossy blue-black upper parts, a reddish-brown throat and forehead and a dark blue band on the breast; the underparts are creamy white. In flight, the adult is easily distinguished from other species by the extremely long tail streamers. In young swallows these are short and therefore often lead to confusion with House Martins (page 184). The Swallow arrives at the breeding grounds in early spring and makes its nest in old buildings which are situated near water. The nest is usually built on a ledge or some other support. It is a cup-shaped structure of mud mixed with grass and straw with a few feathers for the lining; it may be used for several years. The courtship display includes aerial chases (and sometimes fights); the flights are frequently accompanied by a loud song. Normally two broods are raised each year but there may be three. The female incubates the eggs and is fed by the male for about 16 days when the chicks hatch. Swallows are dependent on insects and must migrate south for the winter to secure a continuous supply of food. They migrate in huge flocks and at the beginning of the autumn gather in trees prior to departure, which takes place at night. On their journey they rest during the day and arrive at their winter quarters in October. North American Barn Swallows migrate to South America. European birds travel to central and southern Africa, some as far as South Africa.

Order Passeriformes. Family Hirundinidae, swallows and martins. Common or Barn Swallow (*Hirundo rustica*), 18 cm (7 in) long, found in Europe, Asia and North America, inhabits open country, usually near human communities and water. It nests in buildings, especially farm outbuildings, and also under eaves of houses; it has been known to nest inside houses as well. The nest, made of mud, is reinforced with straw and grass and has a feather lining. Eggs, 1–6, white with reddish-brown spots, are incubated for about 18 days. Young remain in the nest for about 21 days. Diet consists of insects. Close relatives include the Red-rumped Swallow (*Hirundo daurica*), about 15 cm (6 in) long, found in southern parts of Europe and Asia and in Africa, and the Tree Swallow (*Tachycineta bicolor*), 14 cm (5½ in) long, of North America.

183

HOUSE MARTIN

Closely related to the Swallow (page 182) and similar in appearance, the House Martin is a smaller bird with completely white underparts, feathered legs and toes and a white rump that is clearly visible in flight. The tail is deeply forked but lacks the long streamers characteristic of the Swallow. As its name implies, the House Martin lives predominantly near human habitations and makes its nest under the eaves and cornices of buildings; although it is more likely to be found in the countryside it sometimes ventures to the edges of cities. The nest is almost completely covered, with only a small opening about 4 cm (1½ in) wide at the top, and is built of mud which the parents collect from puddles and pools. The male builds the nest and assists in incubating the eggs and feeding the young. Like Swallows, House Martins feed on the wing and also gather in large flocks in the autumn prior to migrating.

Order Passeriformes. Family Hirundinidae. The House Martin (*Delichon urbica*), 13cm (5in) long, found in Europe, Asia and North Africa, and in winter as far south as South Africa, lives in or near human communities. It nests in colonies, the largest usually being found under bridges. It also nests on buildings and sometimes on rock faces. The bowl-shaped nest is plastered against the wall or other structure. It is made from mud and grass. Old nests are reused from year to year, with repairs made each season. Eggs, 4–5, white, are incubated for 14 days. Young leave nest at about 3 weeks. There are two, sometimes three, broods each season. Diet consists of insects.

WHITE OR PIED WAGTAIL

Order Passeriformes. Family Motacillidae, pipits and wagtails. White Wagtail (*Motacilla alba alba*), 18cm (7in) long, found in continental Europe and Asia, is an occasional visitor to the British Isles. Pied Wagtail (*M. alba yarrelli*), 18cm (7in) long, is found in the British Isles and northern France. Both birds inhabit open country near farms and villages and along rivers, lakes and streams. They usually breed near or on buildings. The nest, situated in banks, walls, tree holes, etc., is made of plant materials such as dried leaves and moss. Eggs, 5–6, are whitish with grey spots and flecks. Diet consists of insects and small aquatic animals. This bird frequently acts as a host to the Cuckoo.

Motacilla alba is a species of wagtail comprising two distinct races: the White Wagtail of continental Europe and Asia (illustrated here) and the Pied Wagtail of Britain and Ireland. Sharing similar characteristics and habits, they are distinguished only by the colour of their plumage. In the adult Pied Wagtail the back is black in the males and grey in the females, merging into black on the rump. White Wagtails have pale grey backs and rumps. Both races have black markings on the neck. Wagtails feed on the ground. They may run quickly on their long legs, with the body held horizontally and the long tail held high, or may walk along with their tail wagging up and down. They have a swift, undulating flight and make long curves in the air. The Pied and White Wagtails are frequently seen near water in open country near human communities.

WREN

Order Passeriformes. Family Troglodytidae. Wren (*Troglodytes troglodytes*), 10cm (4in) long, is found in Great Britain, central Europe, parts of Asia and in North America (where it is known as the Winter Wren) in the far north and most westerly mountain regions. In Europe it inhabits a variety of regions from towns to rocky islands. The nest is made in tree stumps, in holes in walls and among branches. It is made of moss with bits of leaves and grass, and lined with feathers. It has a dome and a small entrance at the side. Eggs, 5–7, are white with reddish-brown spots. The diet is restricted solely to insects and spiders, which accounts for the death of many birds during the winter. Wrens do not occur in Australia; the birds there of that name are a species of warbler.

This tiny, round bird is the only species of wren found in the Old World; all other members of the family are New World birds. It is probable that its ancestors made their way across the Bering Strait to settle as far west as Europe. Today it is found in a great variety of habitats and is almost as common and well-known in Britain as the Robin (page 194). The plumage, alike in both sexes, is brown with dark streaks. Flight is fast and straight, usually low over the ground. The Wren is most easily distinguished by the manner in which it holds its short tail in an upright position. The male has a particularly loud warbling song which ends in a trill; the alarm call is usually *tic tic tic.* At the beginning of the breeding season he builds several rounded, domed nests made predominantly of moss, and from these the female selects one which she lines with feathers before laying her eggs. Two broods are raised each year.

HEDGE SPARROW

Some 12 species of Accentors make up the family Prunellidae, small birds that live mainly in the mountainous regions of central Asia. One exception is the tiny Hedge Sparrow, or Dunnock, which is found throughout much of Europe and inhabits lowlands as well as uplands. It takes up residence in town gardens, parks, heaths, moors and woods, almost anywhere in fact where there is low cover in which to build the nest. Although it is generally a timid creature, it feeds on open ground and in the breeding season may be seen making chasing flights in and out of bushes and trees. Similar in size and appearance to the House Sparrow (page 216), hence the common name, the two birds are nevertheless completely unrelated. The plumage of both sexes is alike, the head, neck and underparts being slate-grey and the upper parts brown with dark streaks.

Order Passeriformes. Family Prunellidae. The Hedge Sparrow (*Prunella modularis*), 14cm (5½in) long, found in Europe and central Asia, inhabits woods, mountainous regions, scrub, gardens and parks. It nests near the ground in young trees, hedges and undergrowth. The tidy nest is made of moss and twigs. Eggs, 4–6, pure blue, are incubated by the female for about 14 days; young are fed by both parents and leave the nest when about 11 days old. Diet consists of insects in summer and seeds in winter. Some birds move south in the autumn. A close relative, the Alpine Accentor (*P. collaris*), 18cm (7in) long, is found in mountainous areas of central Europe. It has similar plumage but the streaks are more pronounced and the throat is pale with black spots.

FLYCATCHER

In Europe the best-known and most widely distributed of the flycatchers is the Spotted Flycatcher (illustrated here), a bird with grey-brown plumage, pale underparts streaked with dark brown and dark spots on the top of its head. Females resemble the males. In common with other members of the family it has a wide, flattened bill and short feet that are adapted for perching. It is characteristically seen perched in an upright position on branches and railings, from which it frequently flits out in pursuit of an insect and then returns to its perch. The Pied Flycatcher, another European species, is slightly smaller and has a narrower range. The male usually has black upper parts and is white underneath, with a white patch on the forehead and white wing bars, although the colour of the head and wings may vary during the summer from greyish to brownish. Females are brown, lack the white patch on the head and have narrower wing bars. The Pied Flycatcher is most often seen in woods in hilly country and, unlike the Spotted Flycatcher, does not readily appear in the open. It usually hunts among the branches of trees or sometimes searches for insects on the ground.

More than 350 species of flycatchers, belonging to the family Muscicapidae, are found in the Old World. All of them are small insectivorous birds that inhabit open wooded country such as orchards, large parks and forest margins. None of them occur in the New World, although the common name flycatcher is applied in North America to members of the family Tyrannidae, the tyrant flycatchers. The tyrant flycatchers are an example of the phenomenon that scientists term convergent evolution – the process by which unrelated species in different parts of the world evolve in similar ways to fill the same ecological niche. Species include the Great Crested Flycatcher of eastern North America and the Eastern Kingbird. In Australia two genera of flycatchers (family Muscicapidae) are found: those of the genus *Rhipidura*, the fantails, and the genus *Pachycephala*, the whistlers. In their native country they are commonly called robins.

Order Passeriformes. Family Muscicapidae, Old World Flycatchers. Spotted Flycatcher (*Muscicapa striata*), 14cm (5$\frac{1}{2}$in) long, found in Europe, Asia, and North Africa, inhabits open wooded country. European birds migrate to Africa for the winter. It nests in holes in walls or trees. The nest is built of grass, wool, etc. Eggs, 4–6, are white with red spots. Diet consists of insects, especially flies, and soft fruit in autumn. Pied Flycatcher (*Ficedula hypoleuca*), 13cm (5in) long, is found in northern Europe and Asia and migrates south in the autumn. It inhabits wooded regions, usually in upland areas. It nests in rotting trees, especially in holes made by woodpeckers. The nest is similar to that of the Spotted Flycatcher. Eggs, 5–7, are pale blue. Diet consists of insects. Great Crested Flycatcher (*Myiarchus crinitus*), 22cm (9in) long; Eastern Kingbird (*Tyrannus tyrannus*), 21cm (8$\frac{1}{2}$in) long.

MISTLE THRUSH

Representatives of the large family of thrushes are found in most parts of the world and include such well-known species as the European Robin (page 194) and the Blackbird (page 192). The Mistle Thrush is the largest European thrush and has a wide distribution throughout Europe and Asia. The plumage of both sexes is almost uniformly brown on the upper parts and pale underneath, with dense, dark spots. The Mistle Thrush has a bounding flight and on the wing can be distinguished by the white underneath the wings and the white tail feathers. At one time called the Stormcock for its habit of singing in all weathers, the male usually delivers its song from the highest tree branches. This bird hops or jumps on the ground, picking up worms, insects and their larvae. In autumn it gathers berries, including those of the mistletoe. It is found in many different habitats, from city gardens to upland moors.

Order Passeriformes. Family Turdidae, comprising some 300 species. Found naturally in most parts of the world except New Zealand and remote islands. Mistle Thrush (*Turdus viscivorus*), 27 cm (10½ in) long, inhabits Europe, western Asia and north-west Africa. The nest is made in trees, sometimes on buildings and rock ledges, from stems, moss, grass and earth. Eggs 3–5, buff to pale blue with greyish and reddish markings. Diet consists of insects, worms and berries. New World relatives include the Wood Thrush (*Hylocichla mustelina*), 20 cm (8 in) long, found in eastern regions, usually in deciduous woods. A similar species is the Hermit Thrush (*H. guttata*), 17 cm (7 in) long, found in evergreen forests of North America.

SONG THRUSH

Native to Europe and Asia, the Song Thrush has been introduced into Australia and New Zealand. In Eurasia it has a narrower range than the Mistle Thrush and does not occur in the most southerly parts inhabited by the latter species, although its choice of habitats is as wide and as varied. Sometimes called the Nightingale of the north, this bird has a charming, repetitive song which can be heard for most of the year. It also has a talent for mimicking the calls of other birds. The diet consists of berries and fruit but it is probably best known for its preference for snails which it hammers against hard objects to break the shell. The Song Thrush is smaller than the Mistle Thrush and has orange underwings rather than white and a less densely spotted breast. The courtship display includes chasing flights usually accompanied by loud screaming.

Order Passeriformes. Family Turdidae. Song Thrush (*Turdus philomelos*), 23 cm (9 in) long, is found in Europe and Asia, and is now also successfully established in Australia and New Zealand. It inhabits gardens, woods and open country, and nests in bushes and trees, among ivy on walls, and sometimes even on the ground. The nest is made of moss, grass and twigs, and is plastered inside with mud mixed with saliva. Eggs, 4–6, greenish-blue with black, or sometimes dark brown or grey markings, are incubated for some 13 days; young remain in the nest for about 2 weeks. Two broods are raised each year. Diet consists of insects and their larvae, worms, snails, slugs and berries.

EUROPEAN BLACKBIRD

The cock Blackbird is probably the best-known of all European birds, being easily distinguished from other species both by its appearance and its flute-like song. Its plumage is generally a uniform black and it has a bright yellow bill, although partly albino birds (such as the one illustrated here) are not uncommon. The hen is brown and darkly speckled underneath. Blackbirds are found in every kind of habitat, from forests and gardens to farmland. They characteristically hop over the ground, then stop, cock their heads and listen for worms. Nests are carefully-made structures of grass, moss, twigs and dried leaves, with a layer of damp earth inside which is lined with grass. The hen incubates the eggs but when the young hatch they are fed by both parents. In North America the orange-breasted American Robin is the New World counterpart of the Old World Blackbird.

Order Passeriformes. Family Turdidae. European Blackbird (*Turdus merula*), 25cm (10in) long, is found in Europe, Asia and north-west Africa in almost every kind of habitat from woods to scrubby regions. The nest of twigs, moss and dried leaves with mud or damp earth on the inside and lined with finer grass, is made in trees, bushes, shrubs and occasionally even on the ground. Eggs, 4–6, are greenish-blue with brown or reddish markings. Diet consists of worms, insects, slugs, snails and berries. American Robin (*T. migratorius*), 25cm (10in) long, found from Canada south to Mexico, inhabits woods and gardens. Plumage is dark grey with a white throat and orange-red breast. Occasional wanderers are seen in Britain.

EUROPEAN REDSTART

Order Passeriformes. Family Turdidae. Redstart (*Phoenicurus phoenicurus*), 14cm (5½ in) long, found in Europe, Asia and north-west Africa, inhabits woods, parks, heaths and gardens. It nests in holes and crevices in trees, walls and in nest boxes. The nest is made from grass and leaves, and is lined with feathers and hair. Eggs, 5–8, are blue-green. Chicks hatch after some 2 weeks and leave the nest after about 12 days. Two broods are raised each year. Cuckoos sometimes lay their eggs in the Redstart's nest. Diet consists mainly of insects, especially flies, and berries in the autumn.

Small, slender and distinctive, this member of the thrush family inhabits woodlands and heaths. During the breeding season the male has a black face and throat, white forehead, slate-grey back and wings, and a fiery red breast and tail. (In some areas it is known as the Firetail.) In autumn its plumage is duller with speckled throat and breast. The female is a sombre brown with yellowish underparts. Apart from its appearance, the Redstart may be recognized by the manner in which it quickly flits its tail up and down, especially during courtship. The Redstart is generally unsociable and the male is strongly territorial. As a consequence there are frequent quarrels and fights, especially during the breeding season. Found in Europe, Asia and north-west Africa, the Redstart migrates as far as central Africa in the autumn.

EUROPEAN ROBIN

The plump, round little Robin is predominantly a woodland bird of both coniferous and broadleaved woods and forests, although it is also found in hedgerows. In Britain it inhabits many gardens the year round. Continental birds of northern regions often move south during harsh winters to southern Europe and Africa. Like its close relatives the thrushes, the Robin spends much of its time on the ground searching for worms and grubs, moving along by hopping and stopping at intervals to flick its tail and wings. In the autumn it also takes grains and berries. The usual call is a scolding *tick tick* and the melodious song, performed by the male during the breeding season and also by the female in winter, is delivered from a perch. The courtship display is a modified form of threat display which is enacted at the margins of the Robin's territory when it is threatened by an intruder. The Robin is strongly territorial and will stoutly defend relatively large areas from all other birds. This behaviour has earned it a reputation for courageousness. Erecting the throat and breast feathers to display the bright red plumage, the Robin sways from side to side. These actions are normally sufficient to deter most intruders, although fights are not uncommon. During courtship the display of the breast and the swaying movements are performed by both cock and hen as they face each other; the female indicates acceptance of the male by submissive postures.

The Robin usually makes its nest in hollows under the thick cover of leaves, ivy or some other vegetation, and in nest boxes, although it is also known to choose a variety of strange places and will even nest in houses. The nest is constructed of moss and dead leaves and lined with materials such as animal hair and plant down. The female incubates the five to seven eggs for about two weeks and is fed by the male during this period. The chicks hatch with black down on their head and back and remain in the nest for two weeks. They are fed by both parents. Juvenile Robins have a spotted brown plumage. During the moult in late summer adults, too, appear almost uniformly brown, losing most of their bright red feathers. The European Robin is restricted to Europe and Asia and is not found in the New World. The American Robin belongs to the genus *Turdus*.

Order Passeriformes. Family Turdidae. European Robin (*Erithacus rubecula*), 14cm (5½in) long, is found in Europe and western Asia. It is resident in Great Britain (where it is the national bird) but is partially migrant in other areas. It inhabits coniferous and broadleaved woods and forests, hedgerows and gardens. Plumage in both sexes is the same: olive-brown on the upper parts with a bright red throat and breast. The amount of red varies from one individual to another. It nests in hollows under thick cover, in nest boxes, houses and holes in walls. The nest is made of moss and dead leaves and lined with fine rootlets, animal hairs and plant down. Eggs; 5–7, are white with reddish-brown markings. Diet consists of worms, grubs, berries and some seeds.

NIGHTINGALE

Order Passeriformes. Family Turdidae. Nightingale (*Luscinia megarhynchos*), 17cm (7in) long, is found in southern Europe, Asia and North Africa and migrates south for the winter. It inhabits woods, parks, and gardens with thick undergrowth. The nest is hidden in brambles, and nettles on or just above the ground. It is made of grass, moss and dried leaves. Eggs, 4–6, olive brown, are incubated for about 2 weeks. Sometimes two broods are raised, but more usually one. Diet consists of insects, worms and berries. The bird feeds on the ground, where it hops along like other members of the family. Thrush Nightingale (*L. luscinia*), 17cm (7in) long, found in northern parts of eastern Europe and Asia, inhabits similar areas to those of the Nightingale but damper.

In appearance the Nightingale is unimpressive and rather ordinary but many people believe it has the most beautiful song of any bird. The reddish-brown plumage of both sexes extends over the head and back, the underparts are pale and the broad tail reddish. In spite of its name, the Nightingale sings its clear, pure song during the day as well as at night; in contrast to the sweet song, its call is a harsh *wheet*. Nightingales breed under thick cover in broadleaved or mixed woods, constructing a deep nest of grasses, moss and dead leaves fairly close to the ground. Their breeding range is restricted to the southern parts of Europe and Asia and to North Africa. A close relative, the Thrush Nightingale, is found in north-eastern parts of Eurasia. This bird is a darker brown and has streaks on its breast. Its song, although louder, is not as melodious.

REED WARBLER

The general term reed warbler is applied to a group of birds that inhabit the reed beds surrounding marshes, lakes, rivers and ponds. They are all members of the genus *Acrocephalus*, small birds with brownish plumage found throughout Europe, Asia, North Africa and Australia. The Great Reed Warbler (illustrated here) is the largest member, with a length of 19 centimetres ($7\frac{1}{2}$ inches). In Europe its range is traditionally restricted to the continent although some pairs are spreading to the southern parts of Britain. The underparts are pale and there is a faint white eye stripe. The smaller, but very similar Reed Warbler has a slightly wider range that extends to England and the southern parts of Scandinavia. Both these birds make an intricately woven, basket-shaped nest which is supported by reed stems or the branches of bushes and small trees close to the water's edge.

Order Passeriformes. Family Sylviidae, consisting of some 400 species of true warblers. The term Old World Warbler is somewhat misleading as some members i.e. the gnatcatchers and kinglets, also occur in the New World. Great Reed Warbler (*Acrocephalus arundinaceus*), 19cm ($7\frac{1}{2}$in) long, found in continental Europe, south-eastern and central Asia, and North Africa, inhabits reed beds near open water, and sometimes bushes and small trees. The nest is made of reeds, grasses and plant fibres and lined with flower heads and plant down. It is attached to reed stems over water or over the ground. Eggs, 4–6, are bluish to greenish with dark markings. Reed Warbler (*A. scirpaceus*), 12cm (5in) long, is found in central Asia, Europe and North Africa.

BLACKCAP

Order Passeriformes. Family Sylviidae. Blackcap (*Sylvia atricapilla*), 14cm (5½in) long, found in Europe, western Siberia and North Africa, inhabits woods, open wooded country and parks. Some are resident in their area while others move southwards for the winter. The nest is made among thick cover such as brambles and nettles, and is usually hung like a basket from the lower branches. Eggs, 4–6, are brown to greenish white or grey with dark brown and grey marks. Two broods are raised each year. Diet consists of insects and berries.

The Blackcap is one of the true warblers. The cock is easily distinguished from other members of the family by its glossy black crown; in the female the crown is reddish-brown. The plumage is brown on the back and wings, and a light grey-blue on the breast and head. This is a common bird of woods and woodland margins where there is ample ground cover of nettles and brambles for the bird to make its nest. The more northerly species are partial migrants and arrive at their breeding grounds in early spring. Males have a sweet, loud warbling song and the alarm call is a sharp *tack tack*. The nest is made fairly close to the ground, hung from the lower branches of shrubs and bushes and is a neat cup-shaped structure of plant materials such as grass, plant down and rootlets. Young Blackcaps hatch after an incubation period of about two weeks and may leave the nest after one week if they are disturbed.

WILLOW WARBLER

Order Passeriformes. Family Sylviidae. Willow Warbler (*Phylloscopus trochilus*), 11cm (4¼in) long, found in Europe, Asia and Africa, inhabits woods, open areas with scattered trees and scrub. The nest is made on the ground under plant cover. It is constructed of moss, grass and dead leaves and lined with feathers and hair. Eggs, 5–7, yellowish-white with reddish spots, are incubated for about 13 days. Young fledge after some 19 days. There is usually one brood but occasionally two. Diet consists of insects gathered from branches of trees and bushes. Chiffchaff (*P. collybita*), 11cm (4¼in) long, is found in most parts of Europe and western Asia.

The Willow Warbler belongs to the genus *Phylloscopus*, a group of birds that are sometimes called the leaf warblers. They are all similar in size and appearance, with greenish plumage and pale yellowish underparts. The Willow Warbler is easily confused with the Chiffchaff although it has paler legs, and its song follows a descending scale whereas the latter bird is known, and named, for its *chiff chaff chiff chaff* tune. The Willow Warbler is one of the first birds to arrive in northern Europe in the spring, having wintered in Africa and southern Europe. It makes its nest on the ground among grass and other cover, wherever there are trees and scrub. The nest is a domed, oval structure resembling that of the Wren (page 186) with a round entrance hole at the side. The Chiffchaff's nest is similar but is usually made just above the ground and has a larger, oval-shaped entrance hole.

GREAT TIT

The family Paridae contains around 50 species of titmice and chickadees which are found in Europe, North America, Asia and Africa. The largest and one of the most common of the European species, the Great Tit is found in gardens. It is a frequent visitor to bird tables and often utilizes nest boxes. In more natural habitats it is found mainly in broadleaved woods. The sexes are very similar, both having a black cap, white cheeks, greenish upper parts with a yellow breast and a belly marked with a long black stripe. The young have yellowish cheeks and the cap is a dark brown. Like most tits it is resident in its area and during the winter gathers in groups to search among the branches of trees and shrubs for food, occasionally feeding on the ground. One of the most characteristic of its varied calls is a *teecher teecher teecher* but the female on its nest can be detected by a sudden hissing, scolding alarm call.

Order Passeriformes. Family Paridae. The Great Tit (*Parus major*), 14cm (5½in) long, found in Europe, Asia, Indonesia and north-west Africa, inhabits broadleaved woods, gardens and sometimes hedgerows and scrub. It nests in a wide variety of holes and crevices (trees, banks, etc.) and in nest boxes. The nest is made of moss mixed with grass, fibres and small roots, and is lined with hair or plant down. Eggs, 6–10, are white with reddish-brown spots. There are two broods each season, sometimes three. Young remain in the nest for about 18 days. Diet consists of insects, seeds and fruit. This tit is also adept at opening the tops of milk bottles from which it will steal a drink.

BLUE TIT

The vividly coloured Blue Tit is easily distinguished from other woodland and garden birds: the crown, wings and tail are bright blue and the underparts greenish-yellow. A black collar and eye streak ring the head and the cheeks are white. The young are similar to the adults but duller and have greenish cheeks. Most tits are great acrobats and the Blue Tit is no exception, hanging upside down from branches when feeding. Like the Great Tit it is found in broadleaved woods and sometimes in hedgerows, but in Britain it is also found in cities, areas avoided by its larger relatives. Blue Tits resident on the European continent tend to be more secretive and are usually found only in wooded country away from human communities. A Blue Tit's presence among the foliage can be detected by the shrill *tsee tsee tsit* call. Both male and female take part in courtship display, raising the crown feathers and wings.

Order Passeriformes. Family Paridae. The Blue Tit (*Parus caeruleus*), 11cm (4½in) long, found in Europe, western parts of Asia and in North Africa, inhabits woods, hedgerows, parks and gardens. Its nesting habits are similar to those of the Great Tit, with the nest made in holes, mainly of moss mixed with other plant materials. Eggs, 10–13, white with reddish-brown spots are slightly smaller than those of the Great Tit. Diet consists of insects, their larvae and eggs. In the autumn and winter seeds and berries are also taken.

COAL TIT

Order Passeriformes. Family Paridae. Coal Tit (*Parus ater*), less than 11cm (4½in) long, found in Europe, Asia and North Africa, inhabits lowland and mountainous coniferous woods and lowland broadleaved woods such as oak woods. It nests in holes and in nest boxes. The typical mossy nest is lined with hair. Two broods are raised each year. Young remain in the nest for about 16 days. Diet consists of insects and their larvae, and spiders, berries and nuts. Like the Willow Tit (page 205), the Coal Tit will often hoard some of its food under moss, lichen and bark.

Seen from a distance, the Coal Tit can be confused with the Great Tit (page 200) but it is a much smaller bird with a distinctive and clearly visible white patch running from the back of the crown to the nape of the neck. The chin is black, the underparts are buff and there are two narrow white bars on the wings. Common in coniferous forests, it is also found in upland broadleaved woods. It makes its nest in trees, stumps and stone walls but more often in holes on or near the ground such as the abandoned burrows of rabbits or mice. In summer the Coal Tit feeds on insects and their larvae and also spiders which it hunts among the branches and bark of trees or on the ground. The winter diet consists mainly of nuts and conifer seeds.

CRESTED TIT

The black and white pointed crest on the head makes it impossible to confuse the Crested Tit with other members of the family. As with most tits the sexes are alike, both having grey-brown upper parts and a pale breast. The young resemble their parents but have a shorter and darker crest. Attractive and noisy, its calls are a soft, rolling purr or a sharp *zee zee zee*. The Crested Tit is a forest-dweller that makes its nest in a variety of holes and crevices above the ground. Found mainly in continental Europe, it is extremely rare in Britain where it is mainly confined to pine forests in the highlands of Scotland. Continental birds are also residents of coniferous forests but are sometimes found in broadleaved woods, especially in the western parts of their range. A few wanderers from the continent occasionally make their way to southern England.

Order Passeriformes. Family Paridae. Crested Tit (*Parus cristatus*), 11cm ($4\frac{1}{2}$) long, found in Europe, inhabits coniferous and sometimes broadleaved forests. The nest is made in decaying trees and stumps, in abandoned holes of other birds such as the woodpeckers and in squirrel dreys or wood piles. It is constructed of moss, other plant materials and animal hairs. Eggs, 5–10, are white with reddish-brown spots. Diet consists of insects, their larvae and eggs in the spring and summer, and seeds in the autumn and winter.

LONG-TAILED TIT

The titmice of the genus *Aegithalos* are distinguished from the genus *Parus* mainly by their nesting habits. These birds make oval, dome-shaped structures with an entrance hole at the side. The Long-tailed Tit constructs its nest in the branches of low-growing shrubs, thorns and bramble or in the forks of trees. This tiny, round bird has a tail that is almost the same length as its body, brownish-black upper parts with a delicate reddish tinge on the shoulders and a white breast. The Long-tailed Tit, illustrated here, is found in western Europe, including Britain, and is distinguished by the black stripe running from above the eye to the back of the neck; a northern race has a pure white head. Both races inhabit woodland margins and hedgerows and during the winter perch in great flocks on trees.

Order Passeriformes. Family Paridae. Long-tailed Tit (*Aegithalos caudatus*) is 14cm (5½in) long, including the tail. Found in Europe through central Asia to Japan, it inhabits woodland, hedgerows and scrub but is rarely found in dense forests. The nest is made in tree forks, usually in deciduous trees and among thorn and bramble thickets. The domed nest is constructed of moss, spiders' webs and hair, and is lined with feathers. Eggs, 6–12, are white and marked with fine reddish-brown spots. Two broods are raised each year. Diet consists of insects.

WILLOW TIT

Generally a solitary bird that lives alone or in pairs, the Willow Tit seldom, if ever, keeps company with other members of the same species or with other tits. A relatively rare bird, it lives in damp coniferous forests such as pine and spruce, or in broadleaved woods of oak and birch, and can also be found along rivers edged with willow. A timid bird, it will quickly dive into cover when disturbed, uttering its distinctive *tchay tchay tchay* call. The Willow Tit resembles the Marsh Tit in habits and plumage and it is difficult to tell the two birds apart unless the call and song of each is well-known. Both have black caps or crowns (glossy in the Marsh Tit and matt in the Willow Tit), brownish backs and pale underparts and cheeks. These birds feed on insects and their larvae, and spiders and seeds which they often hoard like the Coal Tit (page 202). In North America the Willow Tit's counterpart is the well-known Black-capped Chickadee.

Order Passeriformes. Family Paridae. Willow Tit (*Parus montanus*), 11cm (4½in) long, found in Europe and Asia, inhabits coniferous and broadleaved woods and copses. The nest is made in a hole in rotten wood such as dead trees and stumps. Wood chips are used to make the nest, which is lined with animal hair and plant heads. Eggs, 6–8, are white with reddish marks. There are usually two broods a year. Marsh Tit (*P. palustris*), 11cm (4½in) long, found in Europe and Asia, inhabits similar habitats to the Willow Tit but avoids fir woods. The nest is similar to the Willow Tit's but is made in natural holes. Eggs, 7–10, are white with reddish-brown spots. Black-capped Chickadee (*P. atricapillus*), 12cm (5in) long, is found in North America.

205

NUTHATCH

Some 15 species of nuthatch belong to the family Sittidae which has representatives in both the Old and New Worlds. Like the woodpeckers they are true climbers but surpass the ability of these and all other birds. They move up and down tree trunks and rock faces supported solely by their feet and are also able to descend head first. Most of them are solitary except in the breeding season and are found in woods and copses in Europe, Asia, Africa, North America and Australia.

The European or Common Nuthatch (illustrated here) inhabits broadleaved woods throughout most of Europe and is also found in Asia. Its plumage is grey-blue with reddish underparts and it has a long black eye stripe, a short tail and a long, straight bill. The bill is strong enough to hammer open the hard shells of seeds and to pry seeds from coniferous cones, an ability which gave the bird its English name which comes from the old English 'nut hacker'. The Nuthatch nests in holes and cavities in trees. Unlike the woodpeckers, it does not excavate the holes but takes over those of other tree-nesters, enlarges natural cavities or sometimes nests in nest boxes. The hen makes a large opening narrower by filling it with clay, possibly as protection against predators. She takes up residence in the nest before she lays her eggs, usually a clutch of six or eight, which are incubated for about two weeks. The young hatch naked and leave the nest after some two weeks.

Five Australian species are known locally as sittellas. In most respects these have similar habits to other nuthatches, although they construct woven nests of bark, spiders' webs and caterpillar cocoons in the branches of trees. Four species are found in North America, the most common being the White-breasted and Red-breasted Nuthatches. The White-breasted Nuthatch frequents broadleaved woods and orchards, while the Red-breasted Nuthatch prefers coniferous forests in northern parts of the United States and in Canada.

Order Passeriformes. Family Sittidae, comprises about 20 species. Common or European Nuthatch (*Sitta europaea*), 14cm (5½in) long, found in Europe and Asia, inhabits broadleaved woods and copses. It nests in holes and cavities, natural or made by other birds, and in nest boxes. The nest is lined with bark and dead leaves. Eggs, 6–9, are white with reddish-brown spots. There is generally only one brood in a year. Diet consists of insects and their larvae, and seeds. New World species: White-breasted Nuthatch (*S. carolinensis*), 14cm (5½in) long, is found in eastern and western regions of North America, in orchards and broadleaved woods; Red-breasted Nuthatch (*S. canadensis*), 11cm (4½in) long, is found in Canada and parts of the northern United States.

BULLFINCH

The Bullfinch is a stout, stocky little bird of woodland margins, hedgerows and gardens. The cock has the brightly coloured plumage typical of the family Fringillidae. Its breast and cheeks are bright red, the back grey, and the head, wing tips and tail are black. Females are pale brown below and brown above rather than grey. On the wing the bird can be identified by its white rump and undulating flight. Although most finches gather in flocks, the Bullfinch is a more solitary bird and is often seen in pairs. Male and female can be seen feeding together throughout the year and are believed to mate for life. Their diet is predominantly vegetarian and consists of fruits, seeds and buds, although the young are sometimes fed on insects. Unlike other finches, which are renowned for their song, the Bullfinch has an undistinguished warble; the call is a soft *tew*, sometimes repeated several times.

Order Passeriformes. Family Fringillidae. Bullfinch (*Pyrrhula pyrrhula*) is 15cm (6in) long. Several races are found from Europe across Asia to Japan. In Europe it inhabits the edges of coniferous and broadleaved woods, hedges and thickets. Usually resident, some birds, however, move further south for the winter. The nest is made under thick cover in trees, shrubs, and ivy and is a cup-shaped structure of fine rootlets and hair on coarser roots and twigs. Eggs, 4–6, pale blue with reddish-brown or black spots. The young are fed on regurgitated food from parents' crops. Diet consists of buds and blossoms in spring, seeds in winter and fruit in the autumn. Because of its habit of attacking fruit trees the Bullfinch is considered a pest by fruit growers.

EUROPEAN GOLDFINCH

Order Passeriformes. Family Fringillidae European Goldfinch (*Carduelis carduelis*), 12cm (5in) long, is found in Europe, north-west Africa and Asia. Most birds are partial migrants and leave their breeding sites in the autumn. It inhabits hedgerows and open areas with trees, and nests in trees, bushes, and hedgerows with thick cover. It makes the nest from twigs, rootlets and fibre from bark and lines it with plant down, especially from thistles. Eggs, 4-6, pale blue with reddish-brown spots. Young fledge at about 2 weeks and are similar to parents but lack the bright red facial mask. Diet consists of seeds and occasionally insects.

The colourful, brightly patterned plumage is characteristic of both male and female Goldfinches. Their song is no less attractive and made them popular cage birds when the vogue for keeping songbirds was at its height. In recent years a ban on trapping birds in Great Britain has resulted in increased numbers there. The European Goldfinch usually lives in close proximity to man, in villages, towns and orchards wherever there is an abundance of composite plants such as dandelions, burdock and particularly thistles. With its narrow, pointed bill, the Goldfinch gathers seeds from flower heads or takes the seeds from alder, birch and other catkin-producing trees. The scientific name – *Carduelis carduelis* – comes from the Latin *carduus*, meaning thistle. A flock of Goldfinches, called a charm, will feed together quite openly with seemingly little fear of intruders. Its colouring makes it easily recognizable.

SERIN

The Serin, a yellowish-green bird with brown streaks and a yellowish breast, belongs to the family of finches. It is found in continental Europe and North Africa and breeds only rarely in the southern parts of England. The Serin frequents forest margins, copses and orchards but it is also found close to human habitations in villages, parks and gardens. It feeds on the ground, hopping along to gather various seeds. During the breeding season several males may fly together in display, making loops and circles in the air. The Serin is closely related to the Wild Canary, a finch that is native to the western Canary Islands, Madeira and the Azores. The ancestor of the well-known domesticated songbird, the Wild Canary was brought to Europe in great numbers by the Spaniards shortly after 1478 and quickly became popular as a captive songbird.

Order Passeriformes. Family Fringillidae. Serin (*Serinus serinus*), 11cm (4¼in) long, found in continental Europe and North Africa, inhabits forest margins, orchards, vineyards, villages, parks and gardens. It nests in trees, bushes and shrubs, usually at least 2 metres (6½ft) from the ground. The nest is made from moss, stems and rootlets, and lined with plant down and animal hair. Eggs, 4–5, bluish-white with reddish and brown spots. Two broods are usually raised each year. Wild Canary (*S. canaria*), found in the western Canary Islands, Madeira and the Azores, inhabits open areas with bushes as well as cultivated land lined with trees. It has also been introduced into Bermuda.

Order Passeriformes. Family Fringillidae. Lesser Redpoll (*Acanthis flammea cabaret*), 12cm (5in) long, found in the British Isles and central Europe, inhabits wooded areas, parks and gardens. The nest is made in bushes, trees and hedges. It is constructed of twigs and moss, and lined with feathers and plant down. Eggs, 4–6, blue with reddish spots. There are one or two broods each season. Diet consists of seeds and insects. Mealy Redpoll (*A. flammea flammea*), 12cm (5in) long, is found in northern Europe. Its habitat and nesting habits are similar to the Lesser Redpoll's. Both birds are closely related to the Goldfinch (page 209). The Arctic Redpoll (*A. hornemanni*), is found in the Arctic regions of Europe, Asıa and North America.

LESSER REDPOLL

The redpolls belong to the family of finches. Several races, including the Lesser Redpoll, are found in parts of Europe, including the British Isles. They generally breed among birch, hawthorn or alder, in coniferous forests and also in parks and gardens where shrubbery is present. The Lesser Redpoll, one of the smallest of the group, is fairly widespread throughout England, Scotland, Ireland and Wales. It has dark brown upper parts, pale streaked underparts and light bars on the wings. In the breeding season the male usually has a bright pink breast. The Mealy Redpoll is most common in northern Europe; it is larger than the Lesser Redpoll and has a paler plumage. Both races feed on seeds which they extract from the catkins of birch and alder. In summer their diet also includes insects.

SISKIN

The Siskin is an active little bird originally confined in Europe to northern and central regions, but recently it has extended its range to England and Wales where some birds have become resident. During the breeding season it is found in large groups in coniferous forests. In winter it feeds near streams and lakes where alder and birch are abundant. The male Siskin is greenish-yellow with a black cap and chin, and pale streaked underparts; the female is duller and lacks the dark cap. When the male displays by flying with outspread wings, the yellowish rump is clearly visible. This characteristic distinguishes it from the redpolls with which it may be confused at a distance. The Siskin makes its nest in tall trees such as spruce, usually some 18 metres (59 feet) from the ground. The nest is constructed of moss, twigs, lichen and cobwebs, and is lined with plant down and feathers. The Siskin feeds mainly on insects during the breeding season and on the seeds of pine, alder and birch during the winter. It is closely related to the Pine Siskin of North America, which lives in similar habitats in both the western and eastern regions of the continent.

The family of finches, of which the Siskin is a member, contains over 125 species found in most parts of the world. One exception is Australia, although several species have been introduced there. Most are tree-dwelling songbirds which feed predominantly on seeds. The males usually have a bright, variegated plumage and in many species the females resemble the males. They are stout birds with sharp, conical beaks adapted for gathering seeds and grains, but many finches feed their young on a diet of insects until they are fledged. This family also includes a group of birds known as Darwin's finches which are found on the Galapagos Islands. They were named after Charles Darwin who studied them while on his voyage on the *Beagle*. These birds evolved from a South American species and now exhibit entirely different feeding habits from their ancestor. The Woodpecker Finch, for example, extracts insects from crevices by using a cactus spine as a tool.

Order Passeriformes. Family Fringillidae. Siskin (*Carduelis spinus*), 12cm (5in) long, found in northern and central Europe, England, Wales and Asia, inhabits coniferous forests in spring and summer and broad-leaved forests in winter. The nest is made in high branches, at least 18 metres (59ft) from the ground. It is cup-shaped and is woven of moss, lichen, bark, twigs and spiders' webs, and is lined with plant down, feathers and some animal hair. Eggs 4–6, very pale blue with reddish and brown spots at one end. Two broods are raised each year. Pine Siskin (*C. pinus*), 12cm (5in) long, found in both eastern and western regions of North America, inhabits coniferous forests and is seen in gardens in winter.

LINNET

Like the Goldfinch (page 209) the Linnet has a beautiful song and was once highly prized as a cage bird, although its colours are not as bright or as variegated. In appearance it is more like the Redpoll (page 211) but it lacks the black marking under the beak. The plumage of the male is dark brown with a crimson cap and breast patch, the head is grey and the underparts grey-brown. The female is similar but lacks the bright red colours of the head and breast. This finch is very common in Europe and is also found in North Africa and parts of Asia. It inhabits woods, hedges, thickets and dunes, especially regions with gorse and evergreen shrubs, although it is not found among dense vegetation. The Linnet often nests in small groups in thickets and bushes and during the winter it forms flocks which often fly together with a wavering, undulating flight. Like the other finches it feeds on the seeds of herbaceous plants.

Order Passeriformes. Family Fringillidae. Linnet (*Acanthis cannabina*), 13cm (5in) long, found in Europe, North Africa and parts of Asia, inhabits woodland, hedges, thickets such as brambles, gorse and bracken, and dunes. The nest is made low in bushes and thickets and in heather or grass. It is constructed of twigs and rootlets, grass, moss and plant fibres and is lined with plant down and animal hair Eggs, 4–6, bluish-white with reddish-brown spots, are incubated for about 2 weeks. Young have yellowish-grey down and remain in the nest for about 2 weeks. Diet consists of seeds.

CHAFFINCH

The commonest finch over much of Europe, during the breeding season the Chaffinch is found wherever there are trees or tall shrubs for nesting. In the autumn and winter it feeds in flocks in beech woods or on farmland, gathering seeds from the ground. The hen is a yellowish-brown but the cock displays a variety of colours: grey-blue, chestnut, muted pink, green and white. The calls and song are also varied, and differ between birds of different areas, but most Chaffinches call the characteristic *pink pink* when alarmed, for which they have been named Spink by country people. The male selects and marks out his territory in early spring and pursues the female with a chasing flight among the trees. Having chosen her nesting site, the hen makes a neat nest some distance from the ground, constructing it from moss and lichen and decorating it with a variety of materials including paper and cobwebs.

Order Passeriformes. Family Fringillidae. Chaffinch (*Fringilla coelebs*), 15cm (6in) long, found in Europe, western Asia and north-west Africa, inhabits farmland, woods, parks and gardens. It nests in trees, bushes and well-wooded hedgerows, at least 2 metres (6½ft) from the ground. The deep, cup-shaped nest is compact and made from moss, lichen, grass and bark fibres, and is lined with feathers and hair. The adult eats seeds and grains and the young are fed small animals such as insects, caterpillars, spiders and earwigs. In winter Chaffinches congregate with other finches and with sparrows in large flocks.

HOUSE SPARROW

Order Passeriformes. Family Ploceidae, true sparrows and weaverbirds. House Sparrow, (*Passer domesticus*), 15cm (6in) long, is found in almost every country in both rural and urban areas. It nests in colonies on buildings and in trees, hedges and nest boxes. The nest is made by both parents and is an untidy structure of scraps, waste material and dried grass. In trees the House Sparrow often builds round nests with a side entrance Eggs. 5–6, white with grey or brown spots. There are three broods a year and sometimes four. Young leave the nest after about two weeks.

Once native to North America, Europe and western Asia, this cheeky and lively little bird is now found in practically every part of the world. It is highly adaptable, and when it has been introduced into other countries it has increased rapidly in numbers and in its range. In North America, especially, (where it is sometimes called the English Sparrow) it has become a common bird in towns, cities and the countryside. The House Sparrow lives in close proximity to man and is usually found wherever there are human communities. It will eat almost anything from scraps to grains, and nests in an amazing variety of places: crooks, crannies, gutters and roofs of houses, and in trees and hedges. Most are resident and their cheerful *chirp chirp* can be heard throughout the year. The cock House Sparrow has a distinctive plumage with a dark grey crown, a black patch from the throat to the breast and a chestnut nape.

TREE SPARROW

The Tree Sparrow and the House Sparrow are the only European representatives of the family Ploceidae; all other members occur in tropical zones from Africa to Australia. The round, domed nest of the sparrow is typical of the weaverbirds, some of which build fantastic structures that provide nesting sites for a number of pairs. The Tree Sparrow is a sociable bird. It nests in colonies but each builds its own nest which is often decorated with bits of bright material. A dark chestnut crown, a clear white patch on the cheeks and a more distinct black throat distinguish this bird from the House Sparrow. The Tree Sparrow does not have the same inclination for living near highly populated areas, and generally prefers open country near arable land. In winter it can be seen in large flocks in fields, often in the company of other birds such as finches to which it is closely related and similar in habitat and appearance.

Order Passeriformes. Family Ploceidae. Tree Sparrow (*Passer montanus*), 14cm (5½in) long, found in Europe, western Asia and North America, inhabits open country near farmbuildings and fields. It nests in holes and crevices in buildings, walls, trees and rocks. The nest is similar to the House Sparrow's, and is lined with animal hair and feathers Eggs, 5–7, whitish with dark spots. Both the House and Tree Sparrow may build more than one nest; one is used during the winter as protection against the cold. Other weaverbird species include the Red-billed Quelea (*Quelea quelea*) from Africa 12cm (5in) long. This bird builds enormous nests in trees which may hold up to 100 brooding females.

Order Passeriformes. Family Sturnidae, starlings and oxpeckers. Starling (*Sturnus vulgaris*), 22cm (8½in) long. Found originally in Europe, it has spread in recent years to Africa and Asia and has also been introduced into other countries, especially the United States. It nests in tree holes, crevices in walls, roofs, buildings, etc. and in nest boxes. The cup-shaped nest is constructed of grass, straw and other plant material and is lined with feathers. It may be decorated with flowers. Eggs, 5–7, glossy blue. Diet consists of worms, insects, snails and slugs as well as fruits, including grapes and olives in warm climates. Close relative, the Hill Mynah or Indian Grackle (*Gracula religiosa*), found in India and south-east Asia, is the well-known mynah bird.

STARLING

The Starling is a gregarious bird that often gathers in large flocks and keeps company with thrushes and crows. It can be seen running jerkily over the ground, searching for worms, snails and insects but it also feeds in rubbish dumps, orchards and vineyards. As an insect and invertebrate eater it benefits gardeners and farmers, although its proclivity for fruit can cause great damage to crops. The Starling will travel long distances in its daily search for food, returning to roost at night in woods or even large city centres. Up to one million Starlings may roost together and in such enormous numbers they can be a nuisance. The plumage of this bird varies considerably from one season to another. In common with other members of the family its feathers wear through the year so that before its annual moult, which takes place in midsummer, its colour appears to be almost black.

RAVEN

Order Passeriformes. Family Corvidae, the crows, magpies and jays. Raven (*Corvus corax*), 64cm (25½in) long, found in Europe, North America and Asia, inhabits mountains, moors and rocky cliffs. It nests on cliff ledges and in crevices and trees, at some distance above the ground. The nest is constructed from sticks and earth and is lined thickly with moss, wool and leaves. Eggs, 4–7, greenish-blue with olive-brown spots, are incubated for about 3 weeks. Young remain in the nest for some 40 days. Ravens probably pair for life and they often use the same nesting site year after year. Most of them are resident where they breed. Diet consists of carrion, birds, eggs, large insects, small mammals, fruits and seeds.

At one time widely distributed throughout most of the northern hemisphere, the Raven is relatively rare in Europe and North America and has become extinct in some areas. It is a typical crow of the family Corvidae and is one of the largest members of the group – some birds grow up to 64 centimetres (25½ inches) long. The tail is wedge-shaped and the plumage is completely black, as are the legs. Close at hand, the bristles covering the nostrils and the long pointed feathers on the throat are apparent. The Raven is found mainly in mountainous country, on rocky cliffs and sometimes on moors. Soaring and wheeling high in the air, the bird utters its characteristic *cronk cronk cronk*. During the breeding season its flight includes intricate displays and diving. Like the vulture (page 139), it is a scavenger and feeds on carrion but it also eats live birds, eggs, small mammals, fruits and seeds.

CARRION CROW

The 100 or more species of crow (Corvidae) are found in Europe, Asia, North America, Australia and Africa. The largest of all the perching birds, they are also considered by many to be the most intelligent. While the Raven (page 219) is held to be a bird of ill-omen, crows generally are a symbol of discord and strife, and figure frequently in many old country sayings. The true crows most common in Europe are the Carrion and Hooded Crows, two races of the species *Corvus corone*. The Hooded Crow, or Hoodie, is found in the eastern and central regions of continental Europe and in the Scottish Highlands and Ireland. Its plumage is ash-grey with black wings, tail and head. Alike in size and behaviour is the black Carrion Crow (illustrated here). It is found in western Europe including England, Wales and southern Scotland. Where their two ranges overlap, the Carrion and Hooded Crows often interbreed, the result of the mating being birds with extremely varied plumage.

The Carrion and Hooded Crows live in almost every type of habitat with the exception of high mountains. In rural areas they are considered great pests, particularly in game parks where they cause great damage by stealing and sucking the eggs of game birds. Their diet, however, is as varied as their habitats and includes carrion, seeds, insects and other small animals. They often feed along shores where they take shellfish and, like the Herring Gull (page 163), drop them from the air to crack the shells open. They can also be seen feeding in rubbish dumps where they compete with gulls for scraps. Contrary to the old belief, these birds are not entirely solitary for they sometimes form large groups when roosting or feeding. Yet one of them perched high in a tree, calling the deep *caw*, or moving in the characteristic direct flight ('as the crow flies') across a field, are familiar sights. The Carrion Crow is often confused with the Rook, a slightly smaller bird with a white cere at the base of the narrower beak and a blue gloss on the black plumage. From a distance, the juveniles especially are difficult to distinguish. The North American Common Crow, found throughout the continent, is very like the European species in both appearance and behaviour.

Order Passeriformes. Family Corvidae. Carrion Crow (*Corvus corone corone*), 47cm (19in) long, is found in western parts of continental Europe, England, Wales and southern Scotland. Hooded Crow (*Corvus corone corvix*), 47cm (19in) long, is found in eastern and central Europe, Ireland and the highlands of Scotland. They nest in tall trees and on rocky cliffs; the Hooded Crow also nests in low bushes. The bowl-shaped nest is made from sticks, twigs and earth and is lined with hair and bark. The Carrion Crow lays 4–6 greenish-blue eggs with dark brown or olive-green marks, and the Hooded Crow lays 5–6 eggs of similar colour. Their diet includes insects, worms, carrion, birds' eggs, berries and fruit. Common or American Crow (*C. brachyrhynchos*). 47cm (19in) long, is found throughout North America. Rook (*C. frugilegus*), 46cm (18in) long, is found in central, northern and eastern Europe (including Britain) and in temperate parts of Asia.

JAY

Some of the most colourful members of the crow family are the jays which are found in both the Old and New Worlds. In Europe the most common species is the Jay, a secretive bird inhabiting broadleaved and coniferous woods. It has a variegated plumage with a conspicuous black tail, white wing patches and rump, and blue and black barred wing covers; when it is excited it erects the black striped crest on its crown. Keen-sighted and alert, the Jay often acts as the watchdog of the woods, warning other birds of intruders with its loud *shaak shaak shaak* scream. The Jay spends most of its time under thick cover but will sometimes feed on the ground where it hops along in search of insects and other small animals. Like the Carrion Crow, it is a menace to gamekeepers, taking eggs and even small birds in the spring. Acorns form the main part of its diet in the winter. When food is scarce in the area it will often forage further south.

Order Passeriformes. Family Corvidae. Jay (*Garrulus glandarius*), 32cm (12½in) long, found in Europe and most parts of Asia with the exception of southerly regions, inhabits coniferous and broadleaved woods It nests in trees and bushes sometimes 2 metres (80in) above the ground. The nest is made from sticks and twigs and is intricately lined with fine rootlets. Eggs, 5–7, greenish-grey or olive brown with dark brown markings. Diet includes birds' eggs, young birds, acorns, other nuts and fruit, grubs and insects. North American species include the Blue Jay (*Cyanocitta cristata*), 30cm (12in) long, found east of the Rocky Mountains in Canada and the United States through to the south.

REPTILES

Reptiles were the first vertebrates able to lead an entirely terrestrial existence. Although they are today a relatively insignificant group, for more than 100 million years they inhabited most of the earth and were the dominant life-form. During this period 16 different orders or groups existed, composed of diverse and numerous species. Of these only four orders survived the mysterious events that dramatically, over a relatively short period, brought about the extinction of thousands of species. The history of their development and decline is a fascinating one and is continually being revised as new fossils are discovered which elucidate previously obscure areas of their evolution. Much of what we know about reptiles, and the birds and mammals that evolved from them, is recent. It was only 150 years ago, for example, that the first fossils of the reptiles called dinosaurs were found in England.

Scientists place all reptiles, both living and extinct, in the Class Reptilia. In spite of their present-day status, as a group they hold an important position in the animal kingdom and for this reason it is desirable to look briefly at their past.

Reptiles of the Past

In the Carboniferous Period, some 345 million years ago, amphibians were numerous on the shores and in the warm, shallow waters of the swamp forests. It was during the latter part of this period that the first primitive reptiles appeared, developing directly from the amphibians. These early reptiles are known as the cotylosaurs, the 'stem reptiles' that were the ancestors of all reptilian forms. The Permian and Triassic Periods that followed saw the development and spread of many diverse species. During the late Triassic Period these included the early lizards, the sea turtles and other marine forms such as the ichthyosaurs and placodonts. Ichthyosaurs were carnivorous, fish-like reptiles with dorsal and ventral fins and a flattened tail. Placodonts were more seal-like and had long necks,

long tails, and limbs shaped like paddles. The rhynchocephalians had also appeared and the mammal-like reptiles which during the Age of the Reptiles gave rise to the early mammals (see page 10). There was also a large and very important group of archosaurs (meaning ruling reptiles) which were to dominate the earth later in the Mesozoic Era. These gave rise to the birds.

The earliest archosaurs were the thecodontians (Order Thecodontia): small, carnivorous reptiles that crawled clumsily over the ground on four limbs. These were the ancestors of the crocodilians (the only group of archosaurs that has survived to the present day), the flying pterosaurs such as *Pteranodon* and *Pterodactylus*, whose large wings were supported by long finger bones enabling them to fly with slow, flapping movements, and the dinosaurs. (It should be noted that the flying reptiles mentioned above were not the ancestors of birds; see page 108.)

The dinosaurs are generally thought of as being huge, predacious reptiles but in fact many of them measured only a few metres in length, and large numbers of them were herbivorous, gentle creatures. Dinosaurs are divided into two orders: the Saurischia, or lizard-hipped dinosaurs, and the Ornithischia, or bird-hipped dinosaurs. The two groups are differentiated mainly on the basis of the structure of the hip bones and other anatomical features.

The Order Saurischia was composed of theropods and sauropodomorphs. The former were bipedal creatures that walked on their hind legs and were predominantly flesh-eaters. One of the most famous is *Tyrannosaurus* ('tyrant lizard'), which had powerful hind legs equipped with sharp claws, terrifyingly sharp teeth and forelimbs that were reduced considerably and had only two digits. The sauropodomorphs included enormous, herbivorous dinosaurs. To this group belong *Brontosaurus*, (whose correct name is *Apatosaurus*), *Diplodocus* and *Brachiosaurus*, some of the largest creatures ever to live on land. *Brachiosaurus* attained a height of 12.6 metres (41 feet) and a weight of 80 tonnes; the largest sauropod weighed about 100 tonnes. These dinosaurs had very long necks, long tails and relatively small heads. The saurischians were abundant by the Jurassic Period, some 180 million years ago. The ornithischians, which evolved later, survived until the end of the Cretaceous Period.

The quadruped ornithischians are divided into four groups. They include the horned dinosaurs (the ceratopsians), the armoured dinosaurs (the ankylosaurs), the plated dinosaurs (the stegosaurs) such as *Stegosaurus*, and the plant-eating, duck-billed and bipedal dinosaurs (the ornithopods) such as *Iguanodon*.

Dinosaurs and other reptiles were able to grow to a greater size and become far more active than their ancestors the amphibians. The reason for this is that the amphibians have a permeable skin through which water is lost as well as taken in from the environment. For these animals, under dry, hot conditions water evaporates rapidly from the body so that it soon becomes dehydrated; but in the moist cool conditions they require, they become sluggish and relatively inactive. Reptiles however do not have this problem. First, their skin is not naked but covered with scales; it is also fairly impermeable. Second, the reptiles of the past had no rivals and therefore did not have to remain small and inconspicuous like the early mammals. Third, by becoming larger they were able to remain warm despite the lack of fur or feathers for insulation, and therefore were able to be more active. The larger an animal is, the more is the bulk of the body further away from the skin, so that it takes longer to cool down than small animals. An animal that weighs approximately half a kilogram (about one pound) for example, will take about five minutes to lose one degree Centigrade, while an animal that weighs 50 tonnes will take about a week to lose the same amount of heat. It is also worth noting that at the peak of the reptiles' development, the temperature of the earth was universally more temperate than it is today. In addition, reptiles can control their temperature to a certain extent by moving into the sun when they want to increase their body temperature and into the shade when they want to decrease it. For these reasons, some scientists now believe that dinosaurs in particular were on their way to becoming warm-blooded.

In the latter part of the Cretaceous Period (about 65 million years ago), not only the dinosaurs but many other land and sea reptiles became extinct. The question why they died out remains a mystery, as does the reason why some of them, few in number compared to their previous relatives, managed to survive until the present.

The living species are grouped into four orders: the Testudines or Chelonia (turtles and tortoises), the Crocodylia or Loricata (crocodiles, alligators, caimans and gharial), the Rhynchocephalia (tuatara) and the Squamata (snakes and lizards).

Living Reptiles
The order Rhynchocephalia contains only one species, the Tuatara (*Sphenodon punctatus*), which is restricted to islands off the coast of New Zealand. This group of reptiles was numerous during the Mesozoic Era and until recently was thought to be extinct. The Tuatara has a lizard-like appearance but anatomically is not as highly evolved. Its body is greenish-grey with spines that extend from the head to the tail along the dorsal surface. In the centre of the head is a pineal eye, an organ that may have sensory functions but is not an optical structure. The Tuatara does not usually grow more than 60 centimetres (24 inches) long. It is nocturnal and completely terrestrial, often living in the burrows of petrels, and is predominantly insectivorous although it also takes small invertebrates and lizards.

Tortoises and turtles (testudines) are easily distinguished from other reptiles by their bony shells which cover the upper and lower surfaces of the body. The upper shell, known as the carapace, is bony and formed by modifications to the ribs. In land-dwelling species it is generally concave and highly arched, while in aquatic species it is more flattened. The lower shell, or plastron, is composed of bony plates. In some species, such as the musk turtles of North America, it is hinged so that the upper and lower sections can be completely closed when the limbs, head and tail are withdrawn into the shell. In testudines or chelonians the sexes are alike in appearance although the males are much smaller and their plastron is generally concave.

Protective armour is also present to a lesser degree in the crocodilians, in the form of scutes lying underneath the scales. In caimans the armour is present on both the upper and lower parts of the body, while in other members it occurs only on the back and tail. Most crocodilians have webbed feet, as do freshwater turtles, but in marine turtles the legs are flattened and broadened to form flippers. Both the testudines and the crocodilians are ancient reptiles that are known to have existed 200 million years ago in the Triassic Period.

The last living order of reptiles, the Squamata, is divided into two large suborders, the lizards (suborder Sauria or Lacertilia) and the snakes (suborder Ophidia or Serpentes). Although lizards are generally regarded as limbed reptiles and the snakes as limbless, in fact both groups contain species that do not conform to this description. The boas for example have rudimentary hind limbs as do some other snakes, and of the lizards the Slow-worm is completely limbless. More reliable distinguishing features are the absence in snakes of eyelids, so that their eyes are constantly open, and external ears. The middle ear is also missing; snakes 'hear' mainly by picking up vibrations from the ground which are transferred from the jaw bones to the skull, although sound waves of very low frequency may also be picked up from the air.

The Class Reptilia contains more venomous animals than any other class of vertebrates. Most of these are snakes, with a few species of lizards also possessing toxic substances. The latter include the Gila Monster (*Heloderma suspectum*) and the Beaded Lizard (*H. horridum*), both large, carnivorous reptiles that have poison glands in the lower jaws. The venom is drawn up into the teeth and is injected into the victim when the lizard bites, causing great pain and often paralysis of the respiratory system.

Venomous snakes are generally divided into three groups on the basis of the construction and position of the fangs. Back-fanged snakes include some members of the family Colubridae, such as the deadly Boomslang. In these snakes (the Opisthoglypha) the fangs are located at the back of the mouth in the upper jaw; the venom from the glands runs down a groove in the fang and from there into the wound. The front-fanged snakes of the group Solenoglypha have large fangs with completely enclosed canals for transmitting the venom. The fangs are so long that when the mouth is closed they must lie against the palate, swinging downwards and forwards when the mouth is opened. These snakes, which include vipers such as the Asp and pit vipers such as the rattlesnake, can usually renew their fangs if they are damaged or lost. The last group, also front-fanged but in which the fangs are immovable, include the mambas, cobras and kraits of the family Elapidae. Spitting cobras belong to the Proteroglypha; these snakes are able to spit their venom over a considerable distance (up to 2 metres, or 6.5 feet) through holes at the front of the fangs. The poison causes extreme irritation of the skin and if it enters the eyes causes considerable damage and even blindness.

Of the two groups of the order Squamata, the lizards are the more ancient, having first appeared about 185 million years ago. The earliest known fossil of a snake probably dates from about 100 million years ago. Snakes, therefore, represent the most modern species among the reptiles.

One of the most important features that distinguishes reptiles from amphibians, as mentioned above, is their fairly waterproof skin which prevents the passage of water both into and out of the body. Reptiles therefore have no need for the mucus-producing glands found in the dermis of amphibians; in fact they possess very few skin glands. Those that do exist are located in specific areas and produce odoriferous secretions used in defence and sexual recognition. The skin itself is dry and covered with scales which arise in the epidermis; a horny outer layer, thin and transparent, is periodically shed or sloughed as the reptile grows. Naturally this occurs more frequently in the rapidly-developing young. The skin may be shed in pieces (as in lizards) or in one piece (as in snakes). Snakes peel off their skin inside out, beginning at the area around the mouth. They rub their face against some rough object, such as a rock, to break the skin and begin the sloughing process.

In many species, especially the lizards, there are crests, spines or frills which arise in the skin and are either modified scales or skin membranes. The horned chameleons, Frilled Lizard and the Moloch or Thorny Devil are examples of reptiles adorned in this way. In rattlesnakes the skin on the tail is loose and segmented and produces the characteristic rattle when the snake vibrates the tail.

All reptiles breathe air through lungs although certain aquatic species can also take in oxygen through the skin surrounding the cloaca and the membranes lining their mouth, thus enabling them to breathe under water.

In common with amphibians, the body temperature of reptiles fluctuates with the temperature of their surroundings. In cold-temperate regions the animal becomes sluggish in the autumn and during the winter it must hibernate as its body metabolism slows down. In deserts and other hot, arid places, many species aestivate during the hot, dry season by burrowing underground. The fact that reptiles are cold-blooded also affects the amount of food they consume for, unlike warm-blooded animals, reptiles do not use food to provide energy for maintaining a constant body temperature.

Reproduction

Although amphibians left the water some 60 million years before the first reptiles appeared, an aquatic environment is still of fundamental importance during the larval and reproductive stages of their lives. For any animal to live in the drier regions of the earth, it is necessary to have a method of reproduction that does not rely on water. It was this feature, evolving first in the reptiles, that helped them to spread and to colonize most of the earth in the Mesozoic Era.

The reptilian egg – and the bird's egg – has a parchment-like shell that is relatively tough and acts as a protective covering. This simple fact has enormous consequences for the breeding habits of reptiles, for it means that the egg can be laid on dry land without the risk this would involve in amphibians of dehydration and death of the embryo. (In amphibians the developing larvae are protected only by a soft, gelatinous substance. They require an external liquid medium to keep them moist.)

The fertilized egg consists of the embryo (initially the germinal disc) which is surrounded by amniotic fluid to keep it wet; the embryo is attached to the yolk, from which it obtains nourishment, by the umbilical peduncle. These structures are completely surrounded by the allantois which collects waste matter from the embryo and also controls the flow of gases that pass in and out of the porous shell and enable the embryo to breathe. Finally there is a thin membrane, the chorion, which lies very close to and is attached to the shell and which has a protective function.

The clutch of eggs is generally laid on the ground (none of the reptiles lays its eggs in water) and left to incubate in the heat of the sun with very little, if any, parental care, although the female python wraps her body round her eggs to keep them warm and protect them. Crocodiles and alligators make some attempt at nesting, as do their relatives the caimans.

Most species are oviparous (the young hatch after the eggs are laid) but some, such as the European Viviparous Lizard and some species of chameleon, are ovoviviparous. In these reptiles the eggs hatch while still inside the mother's body and the young emerge live and active.

SAND LIZARD

Order Squamata, snakes and lizards. Suborder Sauria (Lacertilia). Family Lacertidae. Sand Lizard (*Lacerta agilis*) is usually from 15–18 75cm (6–7½in) long including the tail which is more than half the length of the body. The maximum length recorded is 23.75cm (9¼in). Found in western and central Europe (including England) and in parts of Asia, it inhabits sandy heaths, dunes and rocky regions. It hibernates from September to April. Mating usually takes place in May. At the beginning of June the female lays 5–14 parchment-like eggs in a sandy depression. Young hatch in August and are about 3.75–6.25cm (1½–2½in) long. Diet consists of insects and worms.

The typical lizards belong to the family Lacertidae, of which there are some 150 species in Europe, Asia and Africa. The Sand Lizard exhibits the main characteristics of the group: a slender body with a very long, pointed tail, bony shields on the head and five toes on each limb. Like other lacertids, the Sand Lizard is fond of basking in the sun on rocks and walls. It is extremely agile and disappears into crevices when disturbed, often moving so fast that it is difficult to comprehend its form. This lizard has three rows of dark, white-centred spots running along its body. The females are generally brown or greyish and the males are bright green on the flanks and belly, although the colouring of both sexes may vary considerably according to habitat and age. It is found on sandy heaths, dunes and in rocky areas in western and central Europe and parts of Asia.

EYED LIZARD

Order Squamata. Suborder Sauria (Lacertilia). Family Lacertidae. Eyed Lizard (*Lacerta lepida*) is usually 35cm (14in) long including the tail which is three-quarters of the total length. Maximum length is 75cm (30in). It is found in southern France, Spain, north-west Italy and north-west Africa. Females are generally more brown than the males, with darker spots. Unlike other lacertids, the Eyed Lizard is a ferocious hunter that attacks with lightning speed and bites, shakes and beats its victim until it is stunned or dead. Diet consists of mice, insects, small snakes and other lizards.

The Eyed or Jewelled Lizard gets its name from the blue, black-rimmed spots running in three or four rows along its sides. The back is usually brown with greenish-yellow markings, and the lower portions of the sides and the belly are bright green. This attractive reptile is the largest of the European lizards and the largest of the Lacertidae, measuring up to 75 centimetres (30 inches) long including the tail. The Eyed Lizard often escapes from predators by rapidly climbing a tree but it is perfectly capable of defending itself against dogs and cats by biting their snout. It is equally ferocious with its prey and often when mating; during copulation the male may bite his mate savagely as he grasps her by the neck. The female conceals her six to ten eggs by laying them in tree hollows or burying them in the ground.

Order Squamata. Suborder Sauria (Lacertilia). Family Lacertidae. Wall Lizard (*Lacerta muralis*) is usually 20cm (8in) long including the tail which accounts for two-thirds of the total length. The head is more flattened and the snout more pointed than in other lacertid lizards. The Wall Lizard is found in Europe – in the Mediterranean region, West Germany, Belgium, Holland and France – and the northern parts of Africa and in parts of Asia. It prefers dry, sunny regions and occurs from lowlands to altitudes of 2500 metres (8202ft) above sea level; it inhabits walls, rocky debris and ruins. It hibernates from October to May in the more northerly parts of its range but for a shorter period in the south. Females lay from 3–8 eggs.

WALL LIZARD

This is a relatively small lizard, usually no more than 20 centimetres (8 inches) long, closely related to the Sand and Eyed Lizards. In Europe it occurs most numerously in the Mediterranean region but it is also found further north in West Germany, Holland, France and Belgium. The Wall Lizard is even more agile than other members of the Lacertidae and is able to run rapidly up the steepest walls. Colouring may vary through browns and greys to greens with light-coloured markings, but all forms blend in well with their surroundings. In common with many lizards in this species the tail is readily detached from the body when grasped by a predator, enabling the lizard to escape without being harmed. A new growth appears in place of the lost part, although it never attains the length of the original tail and the scaling is irregular.

SLOW-WORM

The family Anguidae contains species of lizards with either one or two pairs of shortened, reduced limbs or, in the Slow-worm, no legs at all. Indeed the Slow-worm looks more like a snake than a lizard and, like the snake, has a forked tongue but its tail is much longer (longer than its body) and, in common with all other lizards, it has movable eyelids. Adults are generally greyish or brownish with paler underparts. The young are silvery or coppery above and dark below. Slow-worms are predominantly nocturnal and inhabit woods, ditches and grasslands. They are found in Europe (including England, Scotland and Wales), northern Africa and western Asia. Other members of the family include the glass snakes of which there are several species in the United States (where they are known as glass lizards).

Order Squamata. Suborder Sauria (Lacertilia). Family Anguidae – slow-worms and glass snakes. Slow-worm (*Anguis fragilis*) is from 27.5–50 cm (11–20 in) long including the tail. All external evidence of limbs has disappeared although there are vestiges of shoulder and pelvic girdles within the body. Found in Europe, western Asia and northern parts of Africa, it inhabits woods, ditches and grasslands. The female lays from 6 to 20 eggs in August or September and the young hatch immediately. Hibernation occurs from October to April in earth burrows, often in groups of up to 30 individuals. Diet consists of slugs, worms, woodlice and caterpillars. Close relatives include the North American species of glass snakes of the genus *Ophisaurus*.

COMMON IGUANA

Order Squamata. Suborder Sauria (Lacertilia). Family Iguanidae, comprising some 400 species which also include the anolids and the horned lizards. Common or Green Iguana (*Iguana iguana*) measures up to 2 metres (80 in) in length. Found in Central and South America, it inhabits forest edges near rivers and other water bodies; it is also found in mountainous regions and on plains. The female lays from 20–70 eggs which hatch 3 months later. Diet consists predominantly of plant food such as flowers and fruit. This lizard is much hunted in South America and both its eggs and its flesh are eaten by local peoples.

Iguanas are predominantly New World lizards similar in many respects to the Old World agamids (page 232). They occur mainly in the tropical regions of the Americas. Of the two species of true iguana, the Common or Green Iguana of South America is the better known. Generally it inhabits the forest edges bordering rivers, lakes and streams and is an excellent swimmer, diver and climber – climbing trees to a height of 15 metres (49 feet). Both male and female possess a spiny crest running down the back from the head to the tail and an inflatable pouch or sac under the throat; in the male these are used in display to intimidate rivals. When threatened, this iguana will retreat into the trees or water but will also defend itself if escape is impossible, standing its ground and lashing out with its sharp teeth and powerful tail which is equipped with sharp-edged scales. The Common Iguana prefers plant food.

AGAMAS

Order Squamata. Suborder Sauria (Lacertilia). Common Agama (*Agama agama*), about 25cm (10in) long, is found in Africa. Diet includes insects and other invertebrates. The female lays 4–12 eggs. The genus *Agama* is represented in Europe by the Starred Agama (*A. stellio*) which occurs in Greece, parts of Asia and in North Africa. Moloch or Thorny Devil (*Moloch horridus*), about 20cm (8in) long, is found in semi-arid and arid regions of Australia; it feeds exclusively on black ants. The upper parts of the body including the head are completely covered by spiny scales. Frilled Lizard (*Chlamydosaurus kingi*), about 90cm (36in) long, is found in Australia and New Guinea. It has a frill or collar of skin round the neck, erected when threatened.

The family Agamidae contains some 200 species of lizards which are the Old World counterparts of the iguanas (page 231). Most of them occur in south-east Asia and Greece. The Common Agama illustrated here is a true agamid (genus *Agama*) which inhabits rocky, arid regions of Africa, frequently in human communities where it will venture quite boldly into houses. The male is generally brightly coloured with a blue head, an orange throat marked with blue bands and blue or yellow spots on the body. However, like the chameleon and some other lizards it is able to change colour in response to external and internal stimuli. In this species the male is polygamous and will fiercely defend its females from rivals. The genus *Agama* is represented in Europe by the Starred Agama of Greece; close Australian relatives include the Moloch or Thorny Devil and the Frilled Lizard.

CHAMELEONS

Chameleons are truly remarkable lizards, extremely well-adapted to an arboreal, insectivorous life. The body is laterally compressed and has a prehensile tail. The five toes on each limb are grouped in threes and twos, the two groups opposed to each other to form a gripping device. Moving cautiously, as if in slow-motion, the chameleon stalks its prey. The eyes can either move independently of one another, giving the lizard an almost complete picture of what is going on around it, or both can focus on the same object, enabling it to judge distances and so take accurate aim. Once the chameleon is within striking distance it shoots out its long tongue (normally coiled within the mouth) with lightning speed and catches the insect on the sticky, club-shaped tip. Chameleons are probably best-known for their ability to change colour, although many other lizards and animals are also able to do this.

Order Squamata. Suborder Sauria (Lacertilia). Family Chamaeleonidae, comprising some 90 species found in Africa, Malagasy and parts of Europe. Common Chameleon (*Chamaeleo chamaeleon*), found in Spain, Portugal, North Africa and parts of Asia, measures from 22.5–29 cm (9–11½in) long. Diet consists of insects. Female lays from 25–40 eggs in a hollow in the ground. A few species such as the Dwarf Chameleon (*Chamaeleo pumilus*) are ovoviviparous. They are said to give birth to live young but in fact the eggs hatch in the process of being laid. All chameleons are able to change colour. Although in some circumstances they adapt to the colour of their surroundings to some extent, the changes are a reaction to light.

ADDER

The Adder, or Common European Viper, the most widely distributed member of the family Viperidae, is found from Portugal to China and from the Mediterranean to the edge of the Arctic Circle. This snake commonly occurs in dry regions on heaths, moors and at the edges of forests, although it may also inhabit wet or damp areas such as marshlands. The males are generally greyish, the females brownish and both sexes have a zig-zag stripe down the back. However colouring and pattern vary considerably with age, habitat and other factors; there are some forms that are completely black. In common with all other members of the family, the Adder has two long fangs in the upper jaw which pass venom into its prey from glands situated under and behind the eyes. The bite is fatal to small animals on which the Adder feeds. Although it is dangerous and very painful to humans it is not likely to be lethal.

Order Squamata. Suborder Ophidia (Serpentes). Family Viperidae is divided into two groups – the true vipers, which includes the Adder, and the pit vipers. The former are found in Europe, Asia and Africa, the latter in North and South America and Asia. The well-known rattlesnakes belong to this group. Many vipers, including the Adder, are ovoviviparous, i.e. the young hatch from the eggs just as they are being laid. Adder or Common European Viper (*Vipera berus*), usually from 47.5–62.5cm (19–25in) long, has a maximum length of over 80cm (32in). It is found in Europe (including Great Britain) and Asia. Diet consists of mice, birds, rodents, lizards, amphibians and invertebrates.

ASP

Like the Adder, the Asp belongs to the genus *Vipera* and is therefore similar in many respects although it is not as widely distributed. The Asp, or Asp Viper, is found over much of southern Europe, especially in the western and central Mediterranean regions, and occurs north as far as central France and southern Germany. It generally prefers dry habitats such as scrubland and waste ground. The colouring is similar to that of the Adder but there is usually a row of dark spots down the back rather than a zig-zag stripe. The Asp is most easily distinguished from the Adder by the upturned tip of its snout. The fangs are characteristic of the family; they normally lie flat against the roof of the mouth but when the snake is about to attack it opens its mouth and the fangs swing downwards and forwards into the striking position.

Order Squamata. Suborder Ophidia (Serpentes). Family Viperidae. Asp or Asp Viper (*Vipera aspis*) is found in southern Europe. Diet includes mice, young birds and young moles. Mating occurs from March to May, and the female lays from 15–20 eggs. The Asp usually measures from 47.5 to 55 cm (19–22 in) in length, with a maximum length of 60 cm (24 in). Its fangs have enclosed canals which pass venom directly into the wound. In vipers the venom is haemotoxic and affects the circulatory system. One of the deadliest species is Russell's Viper (*V. russelli*). It grows up to 75 cm (30 in) long and is found in India, Sri Lanka and south-east Asia. It is predominantly nocturnal. The diet consists of frogs, rats, mice and birds.

COBRAS

Cobras, of which there are 12 species, are Old World snakes occuring only in Asia and Africa. One of the better known African species is the Egyptian Cobra illustrated here. It was regarded by the ancient Egyptians as a symbol of power and is the snake which is reputed to have brought about Cleopatra's death. Like all cobras it is extremely poisonous but does not have the bad reputation of the Indian Cobra whose aggressiveness has been greatly exaggerated. The Indian Cobra is well-known for its association with snake-charmers who, in reality, exploit the snake's natural behaviour when they appear to mesmerize it with music. When the cobra is disturbed it raises the front part of its body and spreads the neck or hood, to display the 'eye' markings on the back. It also hisses and sways from side to side, thus warning of its intention to strike. These cobras have their fangs broken off to render them harmless.

Order Squamata, Suborder Ophidia (Serpentes). Family Elapidae, comprises kraits, cobras, mambas and coral snakes. Egyptian Cobra (*Naja haje*), about 2 metres (80 in) long, is found in North and East Africa. Female lays 10–12 eggs. Diet consists of small mammals, birds, lizards and frogs. Indian Cobra (*N. naja*), 1.3–1.7 metres (52–68 in) long, is found in central and southern Asia. The largest species of cobra is the King Cobra or Hamadryad (*Ophiophagus hannah*) from 3.75–4.50 metres (12½–15 ft) long, which is also found in Asia. It feeds entirely on other snakes. Relatives of the cobras which are also venomous include the North American coral snakes (*Micrurus spp*) and the Australian Taipan (*Oxyuranus scutellatus*) 4 metres (13½ ft) long.

Order Squamata. Suborder Ophidia (Serpentes). Family Boidae, members of which are most numerous in Central and South America and in the West Indies although they also occur in temperate parts of Asia, Malagasy and parts of Africa. Emerald Boa (*Boa canina*), up to 1.8 metres (6ft) long, is found in South America and inhabits trees. Boa Constrictor (*Constrictor constrictor*), about 1.2 metres (4ft) long, is found in tropical South America. It is mainly terrestrial and inhabits dry areas, taking cover in holes, crevices and hollow trees. Anaconda (*Eunectes murinus*) measures up to 7.5 metres (25ft) long. Found in South America, it inhabits rivers and their banks. The pythons are sometimes regarded as a separate family, the Pythonidae.

EMERALD BOA

This snake is a close relative of the well-known Boa Constrictor and Anaconda. All are members of the family Boidae. Boas have strong, solid teeth but are not venomous; they kill their prey by grasping it firmly in their teeth and coiling their body tightly around the victim until it suffocates. The Emerald Boa of South America is an arboreal species that lives on parrots, monkeys and other tree-dwelling animals. Its body is bright greenish-yellow marked with white rings; the tail is prehensile and acts as a support for the rest of the body while it is wrapped around a branch. Like other boas it is predominantly ovoviviparous and nocturnal. The young of this species are yellowish with dark bands around the body. Other constrictors represented in the Old World are the pythons of Asia, Africa and Australia.

MAMBAS

Order Squamata. Suborder Ophidia (Serpentes). Family Elapidae. Black Mamba (*Dendroaspis polylepis*), up to 4 metres (13½ft) long, is found in dry and desert regions of eastern and southern Africa. It is very aggressive and extremely venomous; one bite contains enough poison to kill five men. Colouring is dark olive green with a light underbelly. Eastern Green Mamba (*D. angusticeps*) and Western Green Mamba (*D. viridus*) are smaller, up to 2 metres (80 in) long, and less aggressive than the Black Mamba. They are found in equatorial and southern Africa, and inhabit trees. Female mambas lay from 9–15 oval, white eggs usually in tree hollows.

Mambas, closely related to the cobras (page 236) and like them extremely poisonous, occur only in Africa. The three species of Green Mamba are all slender and coloured bright emerald green with a yellow belly or yellow markings on the body. They occur in forested areas, often on the edges of rivers, and are entirely arboreal. Coiled around a tree branch, they remain well-hidden among the leaves from where they strike out at their prey – birds, tree lizards and other small animals – with great rapidity and accuracy. The Black Mamba is much longer than its green relatives and is Africa's largest poisonous snake. This species is also much more aggressive and will stand its ground against any predator, with its mouth open and the head raised high in readiness to strike. Unlike the Green Mamba, it is found frequently on the ground where it often makes its home in rock crevices or the deserted holes of other animals.

GRASS SNAKE

Order Squamata. Suborder Ophidia (Serpentes). Family Colubridae comprises some 1800 species found in most parts of the world with the exception of Australia and Antarctica. Grass Snake (*Natrix natrix*), up to 2 metres (80 in) long is found in Europe, North Africa and central Asia. Female lays from 10–30 eggs during July or August in rubbish dumps, compost heaps and other decaying vegetation which provide warmth for the incubating eggs. It hibernates from October to March or April in burrows or among vegetation, sometimes in groups of several individuals. Close North American relatives include the Garter Snake (*Thamnophis sirtalis*), found from Canada to Mexico.

Most of the family Colubridae, including the Grass Snake, are non-venomous species. These snakes are widely distributed in Europe, Asia and North America and are found in trees, on the ground and in watery habitats. The Grass Snake occurs only in the Old World and is common in most parts of continental Europe and in England and Wales. Although its name suggests a terrestrial existence, it is an excellent swimmer and lives in marshland, near rivers and lakes as well as in drier regions. Several races are distinguished, mainly on the basis of colour which varies considerably from a completely black form to an albino form. Generally, however, it is greenish-brown with a yellowish or white collar around the neck and a black and white belly. When handled or disturbed the Grass Snake gives off an offensive odour; it also defends itself by feigning death or making a mock strike by shooting out its head.

Order Squamata. Suborder Ophidia (Serpentes). Family Colubridae. Smooth Snake (*Coronella austriaca*), usually only up to 75cm (30in) in length, is found in Europe, including the southern parts of England (where it is rare), north as far as Scandinavia and east to the Caucasus. It inhabits heaths, scrubland and other dry areas. Diet consists of lizards, insects, and sometimes small mammals. It hibernates from October to April and mates in April; young are born in September. Other colubrids include the egg-eating snakes (*Dasypeltis spp*) of Africa; the tree snakes which include the venomous Boomslang (*Dispholidus typus*) of Africa – a back-fanged snake; and the Hog-nosed Snake (*Heterodon platyrhinos*) of North America.

SMOOTH SNAKE

This snake is found over much of continental Europe and in southern England and inhabits dry rocky or sandy areas such as heaths and scrubland. It belongs to the family Colubridae and like its close relative the Grass Snake (page 239) it varies in colour and pattern and may be brown or grey to yellow or red with dark spots, transverse bands or stripes on the body. The skin is covered with shiny, smooth scales. The Smooth Snake and many other harmless species are often killed by people who mistake them for poisonous species such as the Adder. The female lays from 4 to 15 eggs which hatch immediately into active young. Smooth Snakes feed on lizards such as slow-worms, insects and sometimes small mammals. They may coil themselves around their prey to get a firm grasp.

CROCODILES

The 16 species of true crocodiles are essentially aquatic animals that occur in most of the tropical regions of the world. The long tail and webbed toes facilitate movement through water, and the eyes, ears and nose are placed high on the head and snout, allowing the animal to breathe and observe its surroundings while almost completely submerged. Its jaw is long and powerful. Leathery scales cover the body, and on the back and tail there are large plates (scutes). The tail is laterally compressed and bears a crest of hard plates which makes it a formidable weapon. Dentition is one of the main distinguishing factors between these reptiles. It is the crocodile and not the alligator that is known for its toothy grin. The crocodile gets its expression from the fact that the fourth tooth on each side of the lower jaw protrudes outside the upper jaw when the mouth is closed. Alligators' teeth fit into a pit in the upper jaw.

Order Loricata (Crocodylia) comprises some 25 species of crocodiles, alligators, gavials and caimans. Family Crocodylidae: found in Asia, Africa, the Americas and Australia. Nile Crocodile (*Crocodylus niloticus*) measures up to 4.5 metres (15ft) long. Once very common throughout Africa, it is now extinct in some areas and rare in others. Female lays from 30 to 100 eggs in a hollowed-out nest; eggs are placed in layers, each layer separated by sand and the whole covered over with sand. American Crocodile (*C. acutus*), usually 4.5–6 metres (15–20ft) long, is found from Florida through Central America to Ecuador. It frequently visits the sea. Estuarine or Salt-water Crocodile (*C. porosus*), up to 6 metres (20ft) long, is found from eastern India to Australia.

CAIMANS

Caimans and alligators are found predominantly in North and South America, with one species of alligator occurring in China. They compose the family Alligatoridae and in most respects resemble the crocodiles (page 241) to which they are closely related. Caimans are distinguished from alligators by the broader head, shorter, more rounded snout and more extensive bony armour which covers the belly as well as the back. The five species of caiman are exclusive to South America. In common with many other members of the family, the female caiman builds a nest of vegetation in which she lays her eggs. Plant material is piled into a heap and from 30 to 50 eggs are deposited in a depression made in the top which is then completely covered over. As with other reptiles, the young possess an egg-tooth which they use to break their shell, but they are also aided in leaving the nest by the watchful mother.

Order Loricata (Crocodylia). Family Alligatoridae. Composed of seven species of alligators and caimans. Black Caiman (*Melanosuchus niger*) grows to 3–3.6 metres (10–12ft) long – maximum length 4.5 metres (15ft). Found in the regions of the Amazon and Orinoco river basins, their diet consists of fish, terrestrial mammals and also birds. This species can be dangerous to man and often takes domestic animals such as dogs, sheep and goats. The alligators comprise only two species. Mississippi Alligator (*Alligator mississippiensis*), usually from 2.4–3 metres (8–10ft) long – rarely to 5.7 metres (19ft) long – is found in the south-eastern United States. Chinese Alligator (*A. sinensis*) is found along the Yangtse River.

GREEK TORTOISE

Order Testudines (Chelonia). Family Testudinidae comprises some 140 species of land tortoises found in most warm regions of the world except Australia. The genus *Testudo* contains about 50 of these. Hermann's Tortoise (*Testudo hermanni*), up to 30 cm (12 in) long is found in dry regions along the Mediterranean including Spain, Italy, Greece, Bulgaria and southern Hungary. Greek Tortoise (*T. graeca*) has similar distribution, habits and appearance. Both of these species are brown or green; the plates of the carapace are black with yellow and black edges. They hibernate from November to February, and in the more northerly parts of their range to April. Margined Tortoise (*T. marginata*), is found in southern Greece and Sardinia.

The large group of reptiles known as testudines (or alternatively, chelonians) is composed of tortoises, turtles and terrapins; the name tortoise generally refers to terrestrial species. Of some 140 species of land tortoise, three are sometimes called Greek Tortoises although they are distributed throughout much of southern Europe, particularly along the Mediterranean coasts. As in other members of the group, the carapace (upper shell) is highly arched and the legs are rounded. In Hermann's Tortoise the legs are covered with hard plates and the feet are equipped with long claws. The species actually called the Greek Tortoise has long spurs on each of its legs. These tortoises are vegetarians but occasionally take worms, insects and other small invertebrates. They are frequently kept as pets but many perish because their owners have inadequate knowledge of their hibernating habits and requirements.

243

POND TORTOISE

Order Testudines (Chelonia). Family Emydidae. European Pond Tortoise (*Emys orbicularis*), 25 35cm (10–14in) long, is found in central and southern Europe, north-west Africa and western Asia. Female lays from 9–15 chalky white eggs in a hollow in the ground which she digs out with her hind legs; incubation lasts about 3 months. It hibernates from October to April. Blanding's Turtle (*E. blandingii*) is found mainly in the region of the Great Lakes in North America. Both species have a dark carapace with yellow spots; the newly-hatched young have a tail that is longer than the carapace but the size diminishes gradually with age.

Pond tortoises, which belong to the family Emydidae, form a group between land tortoises and true freshwater turtles, for their habits are both terrestrial and aquatic. Many of them are called terrapins from an Algonquin Indian word meaning 'little turtle', although the word is sometimes also applied, especially in the United States, to some salt-water species. The European Pond Tortoise occurs in central and southern Europe, north-west Africa and western Asia. In common with other members of the family it has a flattened carapace and webbed toes. The European Pond Tortoise is predominantly nocturnal and spends much of its time in the water, although it can be seen during the day basking in the sun. Mainly a carnivorous species, it feeds on frogs, small fishes, snails and worms. Blanding's Turtle is a very close North American relative and in fact is the only other species in the genus *Emys*.

MUSK TURTLES

Musk turtles are small aquatic testudines found only in North America. Both males and females emit an offensive odour when handled or disturbed and have been given the name 'stinkpots' by local people. Apart from coming ashore to lay eggs, the Common Musk Turtle spends most of its time in water. It is not a great swimmer and generally dwells on the bottom of ponds and slow-flowing rivers and streams. In the males the plastron (lower shell) is concave and they are further distinguished from females by horny scales on the hind limbs. The females may lay their hard-shelled eggs in a shallow depression in the earth, covered with sand or vegetable matter, or, as sometimes happens, they may leave them completely exposed on the ground. Other members of the family are the mud turtles which have similar characteristics and habits although, generally, they spend more time on land than the musk turtles.

Order Testudines (Chelonia). Family Kinosternidae, composed of about 20 species of musk and mud turtles. Common Musk Turtle (*Sternotherus odoratus*), from 12 to 14cm (4¾–5½ in) long, is found in eastern Canada and much of the eastern and western parts of the United States. It inhabits ponds, marshes and slow-flowing rivers and streams. Diet consists of small invertebrates such as snails and fishes, plants and carrion. Female lays about 3 eggs. The mud turtles (*Kinosternon spp*) are more widely distributed. In this species, at the front and back the plastron and carapace are joined by connective tissue that can be drawn up so that the turtle is completely enclosed when the head, limbs and tail are withdrawn.

AMPHIBIANS

The Class Amphibia, with some 2400 living species, is a relatively small but diverse group of vertebrates that in most cases begin life in water and later leave to live on land. Usually the larvae bear little resemblance to the adults and the difference in life style between the aquatic and the terrestrial form is considerably marked. It is this feature that gave rise to the name amphibian, which comes from the Greek, meaning double life.

The amphibians are distributed throughout the tropical and warm- to cold-temperate regions of the world; their habitats range from marshes to deserts. The majority have scaleless, naked skin which may be either smooth or rough and warty. In most species the skin consists of two layers, the underlying dermis and a surface layer, the epidermis, which is shed as the amphibian grows and its skin becomes too tight. An amphibian's skin is thin and porous and carries out a number of important functions, one of which is the exchange of gases between the air and the circulatory system so that the amphibian is able to breathe through its skin as well as through its lungs.

Because amphibians are cold-blooded their temperature varies with the temperature of the environment. In cold weather they become torpid and hibernate, usually in mud or earth burrows. In these circumstances the amount of oxygen required for survival decreases and a sufficient supply can be obtained by taking oxygen in through the skin only. In some species, such as the lungless salamanders, respiration is carried out entirely in this manner throughout their life.

The porosity of the skin also enables water to pass both out of and into the body. However, the rate of water exchange varies from one group to another and from one species to the next. Most frogs, for example, lose water fairly rapidly and require damper conditions than do toads, which give off water more slowly, enabling them to tolerate drier areas without the risk of dehydration. An exception to this is found in the water-holding frogs of Australian deserts. They are able to survive dry conditions because they have a large bladder for storing water. In addition they are able to hold water next to their bodies by means of a thin membrane which arises in the epidermis. The membrane completely surrounds the frog and is attached only at the snout.

An amphibian's skin is kept moist by a mucous secretion emitted by glands distributed throughout the dermis. In most species of toads, and in a few frogs and salamanders, the substance is toxic and irritates the mouth and skin of other animals, in some circumstances causing the death of small predators. The most deadly species are the South American poison frogs (genera *Dendrobates* and *Phyllobates*) whose secretions are used by native peoples to make poison for arrows. As in many other amphibians, these frogs have brightly-coloured bodies which act as warning signals to other animals. Colouring comes from three types of pigment cells called chromatophores, each responsible for producing different colours, which are present in the skin. In different species they occur in varying proportions, thus creating the characteristic colour of the species. Different combinations, however, may also occur within the same species, as in the Common Frog where the colouring can vary considerably from one individual to another. In many amphibians the cells may be expanded or contracted in response to external stimuli such as light and internal stimuli such as hormone activity, thus enabling the animal to change colour under certain circumstances.

On the whole, and despite the fact that most are dependent on water for much of their life, amphibians are a highly adaptable group. This adaptation has resulted in great variation in structure, behaviour and appearance. It is therefore convenient to discuss this class in more detail according to the orders to which they have been assigned. The Amphibia have been divided into three main groups: the tailless amphibians (frogs and toads), the tailed amphibians (newts and salamanders) and the blindworms or caecilians.

Frogs and Toads

The tailless amphibians are characterized by a squat body in which the head is continuous with the trunk, i.e. there is no neck. The hind limbs are usually much longer than the front limbs and aid the animal in both swimming and leaping. This group, the largest of the amphibians, contains about 1800 species, most of which are found in damp or wet habitats. They belong to the order Anura (sometimes called the Salientia) and comprise more than 12 families, including the true frogs, tree frogs and true toads.

In the anurans, fertilization of the eggs takes place externally. During courtship the male mounts the female and grasps her with his front legs, either under the armpits or around her loins. This position is known as amplexus. It lasts until the female has deposited all her eggs into the water. As the eggs are laid the male emits his sperm from the cloaca, or vent, and fertilizes them. The eggs are laid in strings or clumps and are protected by a gelatinous substance which is wrapped around each individual egg or the entire mass. The eggs then attach themselves to plants and stones. The newly-hatched larvae or tadpoles

acquire external gills and an elongated tail which gives them mobility so that they can search for food. They have small teeth adapted for rasping and are entirely herbivorous. Over the next several weeks the external gills are replaced by internal ones, the body increases in size and the hind legs appear. Gradually the tail is absorbed, the front legs grow from the area near the gill chamber, and internal changes in the digestive and circulatory systems, as well as the development of lungs, prepare the young toad or frog for its life on land.

This process of metamorphosis and reproduction occurs in the majority of anurans, including the European Common Frog and Common Toad (pages 250 and 255). Generally the parents leave the water after egg-laying and fertilization is complete, leaving the eggs exposed to whatever dangers might befall them. In some species however, one or both of the parents make nests for the eggs or actually care for them themselves. The Gliding Frogs of the genus *Rhacophorus* are tree frogs found in the tropics of south-east Asia. Amplexus and fertilization take place in trees, the female laying the eggs on to a leaf overhanging the water. As they are laid, both parents beat the mucous substance surrounding the eggs with their hind feet, producing a light foam. The larvae develop inside the frothy ball, which hardens on the outside and is watery within. When the crust ruptures or softens, the larvae drop from their arboreal nest to the water below. The males of Darwin's Frog (*Rhinoderma darwini*) have enlarged vocal sacs in which they place the eggs, gathering them up with their tongue. When the eggs hatch, the larvae live in this protected environment while feeding on the egg yolk. The female of the Surinam Toad (*Pipa pipa*) carries the eggs and young in small pockets located in the skin of her back.

Newts and Salamanders
The Order Urodela (or Caudata) contains approximately 150 species of newts and salamanders, divided into seven families. These amphibians have long tails, both in the larval and adult stages. The body is slender and elongated and (unlike the anurans) the four limbs are all about the same size.

Fertilization in these amphibians is generally internal but copulation does not take place. In a few species transfer of the sperm to the female is achieved by apposition of the cloacae, as in some toads; in all others the male's sperm is contained in a capsule known as a spermatophore. This is deposited in the water close to the female, who then moves over the spermatophore and gathers it up with the lips of the cloaca. From this opening the spermatophore enters her body and the sperms, freed from the gelatinous sac, fertilize the eggs. The males generally take on bright colours during the breeding season.

The female lays her eggs either singly or in small groups. They are covered in jelly, as in the anurans. In some species such as the Fire or Spotted Salamander, the eggs hatch inside the female so that the larvae are well-developed by the time they are released into the water. The larvae of newts and salamanders look more like their parents than do the larvae of toads and newts. They are carnivorous and are equipped with sharp teeth. The legs develop fairly quickly with the front ones appearing first. Towards the end of the larval stage the three pairs of external feathery gills disappear and their function is taken over by the newly formed lungs.

Caecilians
The caecilians or blindworms compose the smallest order of amphibians (the Order Apoda or Gymnophiona), with about 70 species in a single family. They live in earth burrows in tropical and warm-temperate regions. Caecilians are seldom seen, and little is known of their reproductive habits except that fertilization is internal.

Caecilians lack limbs and use their heads to burrow in the ground; externally they resemble earthworms, with ringed grooves on the body. They are blind and make use of a tentacle situated near the eye to feel their way about. A second organ is present in the nasal cavity which is used for smelling and tasting.

EUROPEAN COMMON FROG

Order Anura (Salientia), family Ranidae, found in most of the temperate areas of the world except New Zealand and southern Australia. European Common Frog (*Rana temporaria*), up to 10cm (4in) long, hibernates from November to February. Spawning begins at the end of February and the males arrive at the breeding areas before the females. Female lays over 1000 eggs. The male mounts the female and grasps her with his forelimbs, fertilizing the eggs as they are laid. Diet consists of insects, snails and slugs. North American relatives include the American Bull Frog (*R. catesbiana*), up to 20cm (8in) long. It is found in the eastern United States west to Nebraska and has been introduced into California and British Columbia.

Belonging to the family Ranidae, the true frogs, the European Common Frog has a wide distribution throughout most of northern and central Europe, including the British Isles, and ranges east into Asia as far as Japan. In many regions it has become less common mainly because of water pollution and the destruction of its habitats. It inhabits ponds, shallow lakes, marshes and ditches during the breeding season. As soon as the eggs have been laid and fertilized it takes up residence on land among surrounding vegetation or in fields and even gardens. Colouring and pattern vary considerably and may be fawn, yellowish, reddish or dark brown, generally marked with spots. The underparts are whitish or creamy, with grey markings in the males, and brownish or reddish markings in the females. In the breeding season the male develops a growth on the first digit of the front limbs, used to grasp the female.

EDIBLE FROG

The Edible Frog leads a more aquatic existence than its relative the Common Frog, seldom straying far from the water's edge and taking refuge on the muddy bottom when disturbed. The females are larger than the males but both sexes have a light green or brown body marked with darker patches, a white, grey-speckled belly and often a greenish-yellow stripe down the back. Unlike the Common Frog, the male of this species has external vocal pouches at either side of the mouth which inflate or expand as the frog croaks. The Edible Frog inhabits ponds, ditches and brackish waters and does not usually mate until the end of May. This frog is found over much of western Europe, including Italy, Germany, Belgium and France, and has been introduced into southern England. It feeds on worms, small lizards and a variety of insects and their larvae. It often leaps from the water to catch a passing fly, dragonfly or butterfly.

Order Anura (Salientia). Family Ranidae. Edible Frog (*Rana esculenta*). Males are approximately 7.5cm (3in) long, females approximately 9cm (3.6in) long. It is found in central and western Europe and east to western Russia. In this species the hind legs are longer and the webbing of the toes more extensive than in the Common Frog. Diet consists of insects and their larvae, worms, snails and small lizards. The male's loud croak can be heard day and night during the breeding season and summer. Tadpoles may reach a length of 7.5cm (3in); froglets are about 2.5cm (1in) long.

MARSH FROG

This frog is very similar in habits and appearance to the Edible Frog (page 251). It is regarded by some authorities as a subspecies, although it is much larger and inhabits lakes, running water and large ponds. The male's croak resembles laughter and for this reason it is also known as the Laughing Frog. Its colouring is usually olive green to brown, often with the bright green line down the back which may also be present in the Edible Frog. In common with many members of the family Ranidae, the skin is smooth or covered with warts. The breeding season of this species lasts from April through May although the males may be heard until July. It has a varied diet which includes small mammals and occasionally birds.

Order Anura (Salientia). Family Ranidae. Marsh or Laughing Frog (*Rana ridibunda*), from 10–16 cm (4–6⅓ in) long, is found in eastern and south-western Europe, western Asia and north-west Africa. Introduced into south-east England, it is now found in such areas as the Romney Marshes. It inhabits reedy areas of large ponds, lakes and running water and hibernates in mud. Diet consists of insects, newts, other frogs, small mammals and small birds.

AGILE FROG

The Agile Frog is remarkable for its exceptionally long hind legs which are much longer than its body. It can leap a distance of 2 metres (80 in) in a single jump and attain a height of up to 1 metre (40 in). This frog is found in both lowland and upland regions in central Europe, some of the Mediterranean countries and east as far as Turkey and Iran. It is most common in open deciduous woods, particularly beech woods. Breeding takes place in ponds, pools and ditches from late March to late April, occasionally to early May; the eggs are laid in clumps in the deepest parts of the water. Colouring is generally uniform brown but there may be dark speckles. The belly is white or creamy.

Order Anura (Salientia). Family Ranidae. Agile Frog (*Rana dalmatina*). Males are approximately 6cm (2.4in) long, females approximately 8cm (3.2in) long. The male lacks a vocal sac. Found in the Balkans, Spain, France, Germany (infrequently), Turkey and Iran, it inhabits deciduous woods in lowlands up to altitudes of 1300 metres (4264ft). Tadpoles can be up to 6.25cm (2½in) long. The Agile Frog breeds from late March to late April and hibernates in water.

Order Anura (Salientia). Family
Hylidae, comprises about 500
species of tree frog. European
or Green Tree Frog (*Hyla ar-
borea*), up to 5cm (2in) long, is
found in central and southern
Europe, western Asia and north-
west Africa. It breeds from May
to June, usually in clear water
with surrounding thick vegetation. It hibernates in hollow
trees, among tree roots or in
burrows and in summer it inhabits trees and bushes such as
bramble. It feeds on insects.
North American species include
the American Tree Frog (*H. ver-
sicolor*), found from Florida
west to Texas, and the Spring
Peeper (*H. crucifer*) which
occurs in the eastern United
States and south-east Canada.
The Golden Tree Frog (*H. aurea*)
is one of many Australian
species.

TREE FROG

Tree frogs have expanded sticky pads on their feet
which act like suction discs and enable them to climb
with agility. In addition most of the males have a large
vocal sac on the throat. When this is inflated with air it
acts as a resonator for the frog's calls which carry over a
considerable distance. In some species, such as the
Spring Peeper of North America, the call is as melodious
as bird song. The family of tree frogs, Hylidae, is
represented on every continent except Antarctica. It is
most abundant in Australia and South America; only
one species occurs in Europe. In the breeding season the
European or Green Tree Frog (illustrated here) inhabits
meadows, marshes, ponds and lakes, dispersing to
nearby trees and bushes for the summer. The body is
generally green on the upper parts and white on the
underparts, with a fine white and a black line running
from the snout to the end of the body.

COMMON TOAD

Order Anura (Salientia). Family Bufonidae, the true toads, comprises about 340 species native to every continent except Australasia and Antarctica. Common Toad (*Bufo bufo*) grows up to 20 cm (8 in) long; males are smaller than the females. Found in temperate parts of Asia as far east as Japan and in most parts of Europe it is, however, absent from the Mediterranean region and Ireland. It inhabits fields and gardens and migrates to the same breeding ground year after year, travelling along traditional migration routes; breeding begins in mid-March. It is less agile than frogs and slow-moving. A close relative, the Marine or Giant Toad (*B. marinus*), native to tropical America, has been introduced into Australia. American Toad (*B. americanus*).

Most toads are not as dependent on water as their relatives the frogs. They are, therefore, found in fairly dry habitats, migrating to ponds and lakes only during the breeding season. The European Common Toad of the family Bufonidae, the true toads, is a typical representative of the group. Its body is plump, the legs are short and the skin is covered with warts containing glands that secrete a toxic substance. In addition there is a large, cresent-shaped gland behind each eye. The poison causes an unpleasant irritation in the mouth of predators and serves to keep many animals at a distance. Another defensive tactic is the toad's ability to inflate its body, at the same time rising on all fours. The European Common Toad is brown, red or grey with grey to brown specks on a white belly. Outside the breeding season its habits are sedentary and it may occupy the same territory for years.

255

YELLOW-BELLIED TOAD

Predominantly an aquatic species, the Yellow-bellied Toad is found near ponds and in marshes and slow-flowing streams. It takes refuge in water or mud when disturbed. Should it be surprised on dry land, however, it throws itself on its back, displaying its brilliant yellow and black speckled belly. This position, illustrated below, serves as a warning to predators, for the skin glands contain a toxic substance (also found in many other species). Similar behaviour occurs in a closely related species, the Fire-bellied Toad which, as the name indicates, has a brilliant red or orange belly. The Yellow-bellied Toad is found in central and southern Europe where it prefers hilly or mountainous country although it may also occur in lowlands. Its breeding season lasts from April into the summer. The female may spawn several times in one season, attaching her eggs to aquatic plants.

Order Anura (Salientia). Family Discoglossidae; members are distinguished by a rounded, disc-shaped tongue. Yellow-bellied or Mountain Toad (*Bombina variegata*) is 5–6cm (2–2¾in) long. Males are slightly smaller than the females. Found in central and southern Europe, it inhabits watery places such as marshes, ponds, slow-flowing creeks and streams and hibernates on land in sheltered places. The back is olive green to brownish and either plain or speckled. Diet consists of insects and invertebrates such as slugs. Fire-bellied Toad (*B. bombina*), 3.75–5cm (1½–2in) long, is found in central and eastern Europe, inhabiting plains and other low-lying areas. The colouring is greyish-black to dark green on the back.

NATTERJACK

Order Anura (Salientia). Family Bufonidae. Natterjack or Running Toad (*Bufo calamita*) can be from 6–8 cm (2.4–3.2 in) long. Found in Spain, France, the British Isles, most parts of Scandinavia, Poland and western Russia, it inhabits sandy areas, particularly along coasts. It stays near the breeding grounds throughout the year and breeds from April to May (occasionally to the end of June). Egg strings are laid in shallow water and contain up to 4000 eggs. Diet consists of insects and invertebrates.

Remarkable for its very short hind legs, the Natterjack or Running Toad moves more quickly than its relatives, running in short, mouse-like spurts. The body is olive brown to olive green with grey or brown patches; the colour and pattern are rather similar to that of the Green Toad, except for the distinguishing bright yellow line down the middle of the back. The Natterjack is tolerant of both dry areas and brackish waters. It prefers sandy soils where it can burrow a long tunnel and is therefore found especially in coastal dunes and meadows. As in the Common Toad, the eggs are laid in long strings of a gelatinous substance, which in this species may be up to 2 metres (6½ feet) long. It is generally gregarious and may live in groups of several individuals.

MIDWIFE TOAD

The breeding habits of the Midwife Toad are unlike those of other anurans. The female may spawn several times from March to August and lays her eggs on land; the string, which contains only a few large eggs, is immediately gathered up by the male who wraps it around his hind legs and carries it about for several weeks. During the incubation period he seeks shelter in the day and forages at night, also taking the eggs to water to moisten them. After two or three weeks, when the eggs are ready to hatch, he deposits them in the water. The Midwife Toad is native to south-western and central Europe and inhabits hilly country. The colouring is ashy-grey to brownish-grey, sometimes with black or light grey spots; the belly is white and may have greyish-black markings.

Order Anura (Salientia). Family Discoglossidae. Midwife Toad (*Alytes obstetricans*) is from 3–5cm (1⅓–2in) long. Found in Spain, Portugal, France, Belgium, West Germany and Switzerland, it prefers hilly country and conceals itself during the day in wall cavities, under stones and rocks, or in burrows which it excavates with its snout and its front legs. It is sometimes called the Bell Toad because of the bell-like call of the male. It leads a predominantly nocturnal life.

ALPINE NEWT

The Alpine Newt belongs to the family Salamandridae which contains about 33 species of newts and salamanders, many of which are found in Europe. The Alpine Newt, in spite of its name, occurs from central Spain across most of central Europe east as far as Greece, and is found at altitudes of up to 3000 metres (9840 feet). In this species the males are particularly attractive, with a marbled dark grey or bluish back, bright orange or yellowish-orange belly and light blue sides with black spots. Females are generally grey or brownish with dark marbling. Breeding usually begins in late February, although at high altitudes it may be delayed until late summer. The male deposits his sperm in gelatinous capsules known as spermatophores. Moving over the spermatophore, the female gathers it up with the lips of her cloaca so that fertilization of the eggs takes place internally.

Order Urodela (Caudata). Family Salamandridae, comprising 33 species of newts and salamanders. Alpine Newt (*Triturus alpestris*). Males grow to 8 cm ($3\frac{3}{8}$ in) in length, females to 11 cm ($4\frac{1}{2}$ in). Found from Spain through central Europe to northern Greece, it inhabits wooded mountains and breeds in stagnant and running waters. Most of its time is spent on land. Both sexes have a crest along the upper and lower edges of the laterally compressed tail. Males also develop a crest along the back, continuous with the tail crest, during the breeding season. The skin is scaleless and smooth. Female deposits her adhesive eggs on stones and aquatic plants. Close North American relative is the Californian Newt (*Taricha torosa*), up to 17.5 cm (7 in) long.

GREAT CRESTED OR WARTY NEWT

The largest of the European newts, this species spends its life partly on land and partly in the water. Like many of the toads it has warty skin glands which secrete an irritating substance when the amphibian is disturbed. The colour is uniform blackish to brown or may be spotted; the darker sides are covered in small white spots. The belly is orange to yellow and both sexes have crests along both sides of the tail. During the breeding season the male develops a high, toothed crest along the back but it is not continuous with the tail crest. Spawning takes place in March in large lakes, deep ponds or ditches. The male approaches the female, touching head to head, and moves his tail rapidly as he emits the spermatophore so that it is pushed towards the female. As in the Alpine Newt (page 259), the female gathers the spermatophore into her cloaca, where the sperm fertilize the eggs.

Order Urodela (Caudata). Family Salamandridae. Great Crested or Warty Newt (*Triturus cristatus*). Males grow up to 14cm (5½in) long, females up to 16cm (6⅓in) long. Found in continental Europe from central France east to the Ural Mountains, and in England and parts of Scotland and Wales, it inhabits both lowlands and uplands. It hibernates on land in earth burrows and among tree roots. Diet consists of insects, small molluscs such as snails, and the larvae of aquatic insects. Its breeding season begins in March. The female may produce from 200 to 300 eggs. They are adhesive and are deposited singly on to leaves of aquatic plants.

FISHES

Fishes comprise a large and very important group of aquatic animals, for they were the first animals to develop a backbone and so make possible the evolution of all other vertebrate forms (amphibians, reptiles, birds and mammals). The first vertebrates made their appearance some 500 million years ago in the Ordovician Period; 100 million years later fishes were widespread in the Devonian seas. Some of the earliest fishes were the ostracoderms which were very similar to present-day lampreys, the most primitive of living vertebrates. They were bony-plated, jawless fishes that sucked up food through their round mouths. The ancestors of sharks and rays, another primitive group, appeared about 370 million years ago; like lampreys they have a skeleton composed of cartilage. By the Carboniferous Period 300 million years ago the bony fishes, the most highly evolved of the fishes, were very much in evidence.

Bony fishes comprise the largest group of living fishes and are divided into several smaller groups. As well as modern species there are several fishes known as 'living fossils'. These include the air-breathing lungfishes of Australia, South America and Africa, and the lobefin fishes which are thought to have included the ancestors of land vertebrates. Bony fishes, jawless fishes (lampreys and hagfishes) and cartilaginous fishes (sharks and rays), together make up the large superclass fishes.

Form and Structure
Apart from sharks and rays (see pages 263 and 264), the fishes described in the following pages are bony fishes representing the large group of what may be termed typical fishes. Although they exhibit a great deal of diversity and are found in a wide variety of habitats, they are all gill-breathing, cold-blooded vertebrates with spiny or soft-rayed fins that live an exclusively aquatic life. In most the body is symmetrical and fusiform (broad in the middle and tapering at both ends) with fins projecting outwards from the dorsal and anal surfaces and the sides (the pectoral fins). The fins are thin folds of skin supported by rays which generally are jointed at the base so that the fin can be folded and unfolded like a fan. The dorsal, anal and pectoral fins are usually used only to guide and steer the fish, the main propulsion coming from the side to side movement of the caudal, or tail, fin and from the area just in front of the tail known as the caudal peduncle. In addition, many species possess a swim bladder filled with gases that helps them to swim. The swim bladder keeps the body buoyant so that the fish does not have to move the caudal and pectoral fins constantly to keep itself afloat. By varying the amount of gas in the swim bladder the fish can also adjust its buoyancy at different depths.

A lateral line, sometimes called the fish's sixth sense, runs from the head to the tail on both sides of the body. This enables the fish to sense vibrations in the water and so detect objects around it. The line is composed of a row of pores opening into a canal beneath the epidermis which contains tiny organs (neuromasts) that transmit messages to the brain.

Most fishes die of suffocation if they remain out of water for even a short period although some, such as carp, can survive several hours in moist, cool conditions. Apart from the lungfishes mentioned above, fishes cannot utilize oxygen from the air but must take it from the water. Water flows in through the mouth and passes over the gills, thin folds of skin containing thousands of capillaries, or blood vessels, supported by rods of cartilage called gill arches. Oxygen is absorbed by the blood vessels and passes into the blood stream, while carbon dioxide is released into the water. In most bony fishes the gills are protected by gill covers which usually close when the fish takes in a mouthful of water and open as the water flows over the gills.

Reproduction
In fishes the process of reproduction is known as spawning. It may be restricted to a certain

time of year, lasting from a few days to several months, or it may occur throughout the year, such matters generally being related to the habitat in which the fish lives. Fishes in tropical waters, for example, may spawn at any time of the year, while those in cool waters spawn when the water is at its warmest.

Breeding is preceded by the swelling of the reproductive organs and the ripening of eggs in the female and sperm (milt) in the male. At this time the majority of fishes travel to spawning grounds, areas that provide the best conditions for the development of the fertilized eggs and the young. In some species this may involve moving to a quieter area of their habitat; in others it may involve travelling upstream from estuaries or from salt water into freshwater. The European Eel and several species of salmon are examples of fishes that travel thousands of miles to spawn, the former from freshwater to the open sea, the latter from the sea to streams. Most fishes are oviparous; that is, the eggs are released by the female and then fertilized by the male. Some species, such as the Three-spined Stickleback (page 285), lay their eggs in carefully prepared nests. Some lay them among weeds and other objects; many of these species produce eggs that adhere to the plants or rocks. In many cases, however, the fish releases its eggs into open water, either on the surface or on the sea or river bed. Such fishes generally compensate for the hazards the eggs will encounter from predators and changes in temperature by laying thousands of eggs each season. The female Plaice may lay up to half a million eggs in one year.

Bony fishes with copulatory organs, where the eggs are fertilized inside the female, include the guppies (page 278) which give birth to live young.

Habitats

Almost three-quarters of the earth's surface is covered by water, water that provides a great variety of habitats for the world's fishes. These habitats are generally divided into two types: marine, or salt water, and freshwater. Intermediate regions include river estuaries containing both salt and freshwater; these are usually described as brackish waters. Such descriptions however are very broad, for within each type there are areas where widely different conditions prevail. Fishes occurring in the sea for example may be found along shallow, inshore waters (scorpionfishes, wrasses), on the sea bed (rays), among coral reefs in tropical and subtropical waters (butterflyfishes and angelfishes) or in the open sea (the pelagic fishes such as sharks and tunnies). Those found in freshwater may be inhabitants of lowland rivers (breams and roaches), hill streams (graylings and trout) or lakes and ponds (carp and tench). Despite the great diversity and sheer volume of water on the earth however, fishes are not free to wander at will. With few exceptions, most can only survive in waters of a particular temperature, a certain degree of salinity and so on, and therefore are restricted in their range. Many are also tied geographically to an area, especially freshwater fishes that are prohibited by the physical barrier of the seas from travelling from one country or continent to another.

Fishes and Man

Millions of fishes are harvested each year from the sea not only for their flesh, which is a rich source of valuable nutrients, but also for their by-products which are used in the manufacture of fertilizers, medicines and other items. Other fish, such as carp and trout, are artificially bred and raised on fish farms. They are used for food or for stocking ponds and lakes for anglers. Fishes therefore have a considerable commercial value and in addition provide pleasure and recreation for sportsmen and aquarists. Unfortunately, in many regions the wealth of life in seas, rivers and lakes is threatened by man's carelessness. Pollution of the waters by industrial waste, oil slicks and human waste, as well as overfishing, have caused the disappearance of many species from their former habitats and, in extreme cases, the complete destruction of vast areas of coral reefs and plants so necessary for a balanced environment and the survival of all forms of aquatic life.

SHARKS

Class Chondrichthyes, including three Orders: Hexanchiformes, Heterodontiformes and Squaliformes. They are largely marine fishes found in most parts of the world, especially in tropical and subtropical zones. All sharks have an excellent sense of smell but poor sight. The Whale Shark (*Rhincodon typus*) is the largest of all living fishes, and reaches a length of 18 metres (60 ft). Extremely sluggish, it feeds on plankton and small fishes. The Blue Shark (*Prionace glauca*) is found in warm open seas and offshore. It has a bright blue dorsal surface and pale underparts. Diet includes shoaling fishes such as mackerel and herring, and squid. Although this shark has been known to attack man, the ferocity of 'man-eating' sharks has been greatly exaggerated.

Sharks are very ancient fishes found in most of the seas and oceans of the world. They are most numerous in tropical and subtropical regions, with a few occurring in fresh or brackish waters. Many have a torpedo-shaped body, although the Monk or Angel Shark and several others have a flattened form similar to that of the skates and rays (page 264). Unlike bony fishes, the shark has a skeleton composed of cartilage, the gill slits are uncovered and there is no swim bladder to keep it buoyant. The absence of this last feature means the shark must swim constantly to stop itself sinking to the bottom. Its skin is rough and covered with small tooth-like scales (denticles) which are covered with enamel. The teeth grow in rows, one behind the other, and as the front teeth are lost or worn away, others move up from behind to take their place. In most cases they are sharp and pointed for cutting and tearing flesh.

MANTA RAY

Rays, which like the sharks belong to the group of cartilaginous fishes, have enlarged pectoral fins on either side of the body that look very much like wings. In the mantas these are moved up and down in a flapping motion and so propel the fish through the water. The manta or devil ray is distinguished by two fleshy projections called cephalic fins on either side of the head. Its skin is naked but there are one or two spines on the end of the tail. While most rays live on the sea bottom, the mantas are more active and move through all depths of the sea, sometimes making spectacular leaps from the surface into the air. The mouth, on the underside of the head, contains minute teeth, either in one or both jaws. Small plankton, fishes and crustaceans are taken into the mouth with water. The water is then expelled through the gills and the food swallowed. During feeding the cephalic fins are used to guide the food to the mouth.

Order Rajiformes, composed of skates and rays. In these fishes the gill openings are located on the lower surface of the body. Many rays which are bottom-dwellers take water through openings called spiracles (also present in sharks) situated behind each eye. Family Mobulidae, the manta rays or devil rays. European Manta (*Mobula mobular*) is about 1.2 metres (4 feet) wide across the wings, and is found from Portugal to the eastern Mediterranean, and south as far as West Africa. Pacific Manta (*Manta hamiltoni*) is the largest species with a wing span of up to 6 metres (20ft).

TROUT

Class Osteichthyes – bony fishes. Order Salmoniformes. Family Salmonidae. Rainbow Trout (*Salmo gairdneri*) is native to North America, and was introduced into Europe in the 19th century. It can grow up to about 45cm (1½ft) long. It is usually spotted with a pink stripe along the sides from the head to the tail, although colouration may vary. Young take insects and their larvae, while the adults prey on other fishes. Sea-dwelling races are known as steelheads in North America. Brown Trout (*Salmo trutta*), sometimes confused with the Salmon (*S. salar*), has a similar commercial value. Both Brown and Rainbow Trout are frequently raised on fish farms.

These fishes occur naturally in cool and temperate waters of the northern hemisphere, although some have been introduced into regions of the southern hemisphere. Trout prefer cold, fast-moving waters as these contain the high amount of oxygen necessary for their survival. Although many are freshwater species, others spend much of their life in the sea but swim up river to spawn. The young migrate to the sea when they have reached a certain size. One of the better-known species is the Rainbow Trout, a popular sporting fish native to western North America but introduced into many parts of Europe in lakes, rivers and ponds. In North America some populations live in the sea. The species *Salmo trutta* contains several races, known variously as Sea Trout, Lake Trout and River Trout. The Brown Trout is native to Europe but, because of its popularity with game fishermen has been introduced elsewhere.

GRAYLING

Order Salmoniformes. Family Salmonidae, subfamily Thymallinae composed of 1 genus and 6 species. European Grayling (*Thymallus thymallus*) grows to about 50cm (20in) long, with a weight of up to 1.5kg (3½lb). Found in many parts of Europe, particularly the northern regions, it is, however, absent from Scotland and Ireland. During the spawning season from March until May in the upper reaches of rivers, the males mark out a territory which they defend against rival males. Intricate courtship ritual includes the male inviting the female to spawn in a depression in the gravel, the display of the male's dorsal fin and vibrating movements made by both sexes. Female lays the eggs several times in the same nest, and young hatch in 20–30 days.

Some six species of Grayling are known, all belonging to the genus *Thymallus*, so named because their flesh has an odour very similar to that of thyme. A freshwater fish, it is found in Europe, Asia and North America, generally in cool waters of rivers but occasionally in estuaries and lakes. Like trout (page 265), it requires water that is rich in oxygen. The European Grayling occurs in Britain and the northern and central regions of the continent but it is absent from Ireland and from the northern parts of Scandinavia. It is a greenish-blue fish with large scales, a brightly-coloured, high dorsal fin and a small andipose fin. This fish feeds principally on insects, particularly the mosquito, but large specimens will also take small fishes. Graylings become fully mature at about 4 years and may live for up to 15 years.

PIKE

Order Salmoniformes. Family Esocidae, composed of 5 species of pike, only one of which is found in Europe. Pike (*Esox lucius*) is found in temperate regions of Europe (absent from Spain), Asia and North America. It inhabits clear, weedy freshwaters and sometimes brackish waters. The record catch measured about 150 cm (60 in) and weighed about 35 kg (77 lb), while the average pike measures 45 cm (1½ ft) and weighs 5 kg (11 lb). Females spawn from February to May; a large female may lay 200,000 eggs a year. Colouring is grey-green or brownish on the dorsal surface with yellowish spots on a green background on the sides; underparts are pale.

Solitary, carnivorous and aggressive, the Pike is one of the most highly-prized fish among anglers for, in addition to its sporting value, it is an excellent table fish. It has a slender body and an elongated head with a large mouth and numerous teeth. Both the anal and dorsal fins are set very near the caudal fin which is forked. Found throughout the major regions of the northern hemisphere, in most parts of Europe, in Asia and North America (where it is known as the Northern Pike), it inhabits the clear, weedy waters of lakes and slow-running rivers but it is sometimes also found in brackish waters. Unlike the Grayling, the female does not build a nest but scatters her eggs in shallow waters or in the flood waters on river banks. The eggs are adhesive and stick to leaves and stems, as do the larvae for a period. The fry (young) feed on invertebrates at first and then on other fishes and vertebrates.

PIRANHA

The piranha belongs to the family Characidae, the characins, which contains some of the most colourful and popular aquarium fishes, including the Neon Tetra. The piranha is also kept in aquaria but because of its ferocity it is unsuitable for the home aquarium. Of more than 15 species, the most common is the Red-bellied Piranha; the largest, *Serrasalmus piraya*, which may grow up to 60 centimetres (24 inches) long. In the wild, piranhas are found in South and Central America, particularly in the Amazon River basin. All of them have numerous, razor-sharp teeth that slice away flesh from prey – normally other fishes but also any large animal (including man) that happens to be in the vicinity of a large shoal. They are attracted by movement in the water and, with their acute sense of smell, by the scent of blood.

Order Cypriniformes. Family Characidae – characins. Found in warm waters from the southern United States through Central America to most of South America. Most are carnivorous although some species with flattened teeth are herbivorous. Genus *Serrasalmus*, containing about 16 species of piranha, is found in South America. Many are brightly coloured and all occur in freshwater. Red-bellied Piranha (*S. nattereri*) measures 15–30 cm (6–12 in) long and is found in the rivers of northern South America, especially in the Amazon River basin. It feeds on other fishes (including members of its own species) and large animals up to the size of a cow. Female spawns on aquatic plants; the eggs are then fertilized with milt.

GOLDFISH

Most of the world's freshwater fishes belong to the Order Cypriniformes including minnows, the many species of carp, and the goldfishes, all of which compose the family Cyprinidae. Cyprinids occur naturally in Europe, North America, Africa and Asia but many species have been introduced into Australia and South America. Both the carp and the goldfish have been associated with man since ancient times. The carp is artificially raised for food and the goldfish is bred as an ornamental fish. The aquarium goldfish varies enormously in appearance, ranging in colour from rich black to bright red, yellow or orange, with some spotted or mottled specimens. Some, such as the Veiltails, have elaborate, delicate, filmy fins and others such as the Celestial Goldfish, have bulging, grotesque faces.

Order Cypriniformes. Family Cyprinidae, composed of some 200 species of freshwater fishes native to most countries of the world with the exception of Australia and the South American countries. Goldfish (*Carassius auratus auratus*), native to the Far East, is today one of the most common aquarium fish, and is found in all parts of the world. Length varies from 15–45 cm (6–18 in). During the breeding season the male pursues the female, stimulating her to spawn. Diet consists of plant material, insect larvae, shrimps and small aquatic animals. Large specimens and those kept in large ponds often revert to the greenish, silvery-grey or black colours of wild species. Common varieties are similar to the golden form of the Gibel Carp, also called the Goldfish.

Order Cypriniformes. Family Cyprinidae. Orfe or Ide (*Leuciscus idus*) grows to 30–50 cm (12–20 in) in length and rarely to 100 cm (40 in). The average weight is 1.5 kg (3½ lb). Found in central and eastern Europe and in parts of Scandinavia and Asia, it inhabits rivers, lakes and occasionally brackish waters. Female lays between 40,000 and 110,000 eggs which hatch in 2 to 3 weeks. Young mature at about 6 years. The relatively deep, laterally compressed body has large scales. It is of commercial value in some countries (i.e. Russia) and in others is a valuable sporting fish.

ORFE

The Orfe or Ide is absent from most waters of southern and western Europe (although it is found in southern England) but is a popular angling fish in central and eastern Europe and is of great economic value in Russia where it is often caught in nets. Predominantly a river fish, it is also found in lakes and occasionally in brackish waters. The Orfe has a relatively long, flattened body, greyish-green above the lateral line and silvery below. The paired fins are reddish, and the remaining fins usually greyish-green. The Golden Orfe is a variety rarely found in the wild but it is bred for use in aquaria and ornamental ponds and pools. This fish may be silvery or orange with a reddish back and light red fins. The Orfe spawns near river banks from March until May, each female laying up to about 100,000 sticky eggs which adhere to plants and other objects.

TENCH

The Tench inhabits the lower reaches of slow-moving rivers and weedy lakes, and occasionally brackish waters. It is able to tolerate waters with a low oxygen level. In all habitats it lives fairly close to the bottom; in very cold or very warm temperatures it buries itself in the mud and ceases to feed. Its colouring is greenish with a gold or coppery sheen. The small scales are deeply set in the skin, which is covered with a heavy slime. It has small eyes, barbels at each corner of the mouth and rounded fins. The Tench, native to Europe and Asia, and introduced into New Zealand, Australia and North America, is often raised on carp farms. As in the Orfe, there is a golden variety bred as an aquarium and ornamental pond fish. In the wild this fish provides good sport for anglers as it is a strong fighter.

Order Cypriniformes. Family Cyprinidae. Tench (*Tinca tinca*) is usually 20–40 cm (8–16 in) long, but may grow up to 60 cm (24 in) long. Its weight is usually between 2.7–3.6 kg (6–8 lb), with a recorded maximum of 7.2 kg (15¾ lb). It is found in most parts of Europe, although it is absent from Scotland and northern parts of Scandinavia; in Asia excluding Russia; and has been introduced into Australia, New Zealand and North America. It inhabits the low reaches of rivers, weedy lakes and estuaries. Female spawns from May until July, laying up to 800,000 eggs a year. The young are mature at about 3 years. Diet consists of invertebrates such as insect larvae, molluscs and crustaceans.

ROACH

The Roach is an omnivorous fish that feeds on aquatic plants and invertebrates and, in large specimens, on molluscs. It often associates with the Common Bream (see page 275) in slow-flowing rivers, lakes, canals and reservoirs, and is sometimes found in brackish waters. In all habitats it lives in large shoals. Several species occur in Europe. These include the Adriatic Roach, the Danube Roach and the Baltic Roach. Species found in saline conditions migrate up river to spawn. The breeding season lasts from May until June and spawning takes place in weedy, shallow waters. The Roach has considerable commercial value as a food fish and is sometimes used by anglers as bait for large carnivorous fish.

Order Cypriniformes. Family Cyprinidae. Roach (*Rutilus rutilus*) is usually 20–35cm (8–14in) long, with a maximum length of 44cm (17in) and a maximum weight of about 2kg (4.4lb). It is found in most parts of Europe (with the exception of Scotland, the Iberian Peninsula and the most northerly regions) and in most parts of Asia. It inhabits lakes, canals, the lower regions of slow-flowing rivers and reservoirs. Spawning occurs from May until June in weedy, standing waters; the female may carry up to 200,000 eggs each year. Larvae and young move about in shoals, as do adults. As breeding is prolific, in small areas where there is an insufficient number of predators conditions become crowded and the Roach may never achieve the usual size.

CARP

Order Cypriniformes. Family Cyprinidae. Carp (*Cyprinus carpio*) usually measures 20–40 cm (8–16 in) long, and rarely up to 1 metre (40 in) long, with a maximum weight of 30 kg (66 lb). It is found in most parts of Europe and was introduced into the United States, New Zealand and Australia. It inhabits lakes, ponds and rivers, and is of great economic importance in Europe where it is raised on fish farms. The female usually spawns in weedy, shallow waters and lays about 300,000 eggs. Young are mature at about 3 years. It is prolific and adaptable. In the United States, where it is virtually free of natural predators, it has spread rapidly and is sometimes considered a pest. Diet consists of plant materials and benthic animals such as molluscs and crustaceans.

The Mirror Carp, illustrated here, is a domesticated or cultivated form of the Carp. Its mirror-like scales vary in size and are irregularly distributed over the body but are usually most abundant along the lateral line. Another form, the Leather Carp, is completely scaleless. The Carp originated in areas of eastern Europe and western Asia but has been known throughout Europe for thousands of years. It is found in lakes and slow-flowing rivers in most parts of the continent and the British Isles. More recently it has been introduced into other countries, including the United States, Australia and New Zealand. Like the Gibel Carp (see page 274), it is very adaptable and can tolerate muddy and even polluted waters. The Carp has large scales covering the entire body and is distinguished by two pairs of barbels, one long and one short. The barbels aid the fish in its search for food which consists of both plants and animals.

GIBEL CARP

The Gibel Carp is also known as the Wild Carp or Goldfish but should not be confused with the Carp (page 273). It is closely related to the popular aquarium goldfishes (see page 269), which the golden and red varieties of the Gibel Carp resemble. Apart from these bright varieties, some Gibel Carp are silvery or brownish. Like the Carp, this fish is extremely adaptable and can survive in weedy and muddy waters. It dwells on the bottom and often associates with the Crucian Carp with which it may crossbreed. The Gibel Carp originated in Asia but has spread westwards into central and north-western Europe and is also found in parts of the British Isles. Females spawn in May or June, laying about 200,000 eggs, some of which may be fertilized by the Carp or Crucian Carp. The offspring that are produced in this way are generally infertile females.

Order Cypriniformes. Family Cyprinidae. Gibel Carp or Wild Carp (*Carassius auratus gibelio*) is usually 25cm (10in) long but may be up to 50cm (20in) long and weighs up to 1.2kg (3lb). It is found in England, north-western and central Europe and in Asia. It commonly has a dark back, yellowish-brown sides and a pale belly, but varies from red or orange to silver and brown. Diet includes plant materials, insect larvae and freshwater shrimps. Crucian Carp (*Carassius carassius*) is about 20–40cm (8–16in) long. Found in slow-flowing rivers and lakes in central and eastern Europe, it was introduced into England and other countries. With a dark back and yellow or greenish sides, it resembles the Carp (but lacks barbels).

COMMON BREAM

The Common Bream is found in similar habitats to that of the Roach (page 272) – slow-flowing rivers, lakes and reservoirs. These two fishes have several characteristics in common. Like the Roach, the Common Bream prefers the river bed where there is an abundance of aquatic vegetation; it generally lives in large shoals and can survive in brackish waters, although in such conditions it migrates upstream to spawn. The Common Bream has a deep, laterally flattened body. Its mouth bends downwards and can be protruded to gather worms, insect larvae and molluscs from the mud. During the spawning season, from May until July, the male has tubercules (pearl organs) on its body and head. The female spawns in shallow waters on aquatic plants and other objects.

Order Cypriniformes. Family Cyprinidae. Common Bream (*Abramis brama*), is usually 30–50 cm (12–20 in) long, sometimes up to 80 cm (32 in) long, and weighs about 5 kg (11 lb). As with the Roach, overcrowded conditions usually result in stunted growth. Found in most parts of Europe, except in the most northerly and southerly regions, and in parts of Asia, it inhabits slow-flowing rivers, lakes and reservoirs; it is also occasionally found in brackish waters and estuaries. It feeds on the river and lake bottom on benthic animals such as worms, molluscs and insect larvae. The female may lay up to 350,000 eggs a year. Young mature after 5 years.

MORAY

Order Anguilliformes, eels. Family Muraenidae, composed of some 120 species. Found in tropical and subtropical seas and in some of the warmer waters of the temperate regions, it inhabits coral reefs and rocky areas of the sea. The skin is thick and leathery to the touch. Eels usually grow to 1.8 metres (6ft) long, and infrequently up to 2.5 metres (8 feet) in length. A few species attain a length of only 15cm (6in). Diet consists of cuttlefish, fishes and crabs. It also feeds on dead fish and other aquatic animals. Green Moray (*Gymnothorax funebris*) is found in the tropical and subtropical waters of North and South America. Zebra Moray (*Echidna zebra*) is a striped species found in the Indian and Pacific Oceans.

Some 20 families make up the group of eels, unique fishes with a long, serpentine body and usually scaleless skin. The pelvic fins are absent and the remaining fins are spineless and joined together so that they appear as one. With only a few exceptions all are marine fishes which occur in most of the waters of the world. Morays are found mainly in subtropical and tropical waters among coral reefs and rocky regions of the sea bed, where they lie in wait for passing prey. They are distinguished from other eels by their large, extremely sharp teeth. Bad-tempered and vicious, they are capable of inflicting deep wounds on humans, in some cases causing death. Well-known species include the Green Moray which occurs in warm waters of North and South America and the Zebra Moray of the Indian and Pacific Oceans.

EEL

Order Anguilliformes. Family Anguillidae, composed of some 16 species found in Europe, North America, South America, Asia and Australia. All species breed in the sea and then travel to freshwaters. European Eel (*Anguilla anguilla*); females grow up to 90 cm (36 in) long, and the males grow up to 30 cm (12 in) long. They are found in freshwater habitats of Europe though some males remain in brackish waters and along the coasts. Females may lay up to 20 million eggs. Diet consists of invertebrates and fishes. American Eel (sometimes considered as a subspecies of *A. anguilla*, otherwise classified as *A. rostrata*) breeds in the Sargasso Sea and takes up to a year to travel approximately 1609 km (1000 miles).

Eels of the family Anguillidae include the European and the American Eels, both of which are called freshwater species although they begin life and spawn in the Atlantic Ocean. Like the salmon, they have a fascinating reproductive cycle which, in the case of the European Eel involves migrating thousands of miles from the spawning grounds to the rivers and lakes where they spend most of their lives. Laying and fertilization of the eggs take place in the Sargasso Sea in the north Atlantic. The eggs hatch into tiny, transparent larvae called *leptocephali* or glass fishes, which then drift on the current and swim towards Europe, a journey that may take up to three years to complete. By the time they have reached their destination they have become small eels, or elvers, and most of them travel upstream to river headwaters and lakes. When fully mature, the adults return to the Atlantic to spawn.

GUPPY

The Guppy belongs to the family Poeciliidae, a group of small fishes that give birth to live young. It is found mainly in warm inland waters of Central and South America and off the coasts of South America and the West Indies, but has been introduced into other subtropical and tropical waters to control mosquito larvae. Guppies are often kept by aquarists and are almost as popular as goldfishes. The males are often brightly coloured, especially on the fins which vary considerably in size and shape. Breeders concentrate on this feature to produce an even greater diversity of forms than those found in the wild. These fishes are extremely prolific, one female producing from 30 to 50 young several times a year; mature females may bear up to 100 young at a time. The male is particularly attentive to the female but neither parent shows any concern for its young. In fact these fishes often eat their offspring after birth.

Order Atheriniformes. Family Poeciliidae, composed of about 18 species found only in the warm waters of North and South America. Apart from the guppies, they include other popular aquarium fishes such as the platys and mollies. Guppy (*Poecilia reticulata*) males grow to about 3cm (1.2in) long and females to about 6cm (2.5in) long. They are found in Central and South America and in the waters of the West Indies. The young are able to swim freely and search for food immediately after birth. They hide from the adults among aquatic plants. Diet is omnivorous.

SEAHORSE

Order Gasterosteiformes. Family Syngnathidae, comprising the pipefishes and seahorses. Pacific Seahorses (*Hippocampus ingens*), the largest species, grows up to 30cm (12in) long. It is found from California to Peru. The smallest, the Dwarf Seahorse (*H. zosterae*), is less than 4cm (1.6in) long and is found in the Atlantic Ocean off the coast of Florida and in the Gulf of Mexico. Two species are found in the Mediterranean Sea. As the movement of seahorses is slow, they have a very large swim bladder to keep them afloat. They have no teeth. The long, tubular snout is used to suck up prey such as small fishes.

One of the most easily recognized of the fishes, the seahorse is found mainly in shallow waters in tropical and subtropical regions but occasionally occurs in temperate zones. Unlike other fishes it swims with its body held vertically and the head bent at an angle; it is propelled slowly and gracefully along by the vibrating movements of the dorsal and pectoral fins. The former drives it forwards and the latter acts mainly as a rudder. Seahorses have a prehensile tail with which they grasp hold of the stems of aquatic plants such as seaweed and can therefore remain hidden among vegetation. They lack a caudal fin and the scales, which are minute, form hard rings that encircle the body. Once the female has laid her eggs she has completed her part in the reproductive process, for she lays them in a brood pouch located on the male's belly. He carries them about with him until the young are fully developed.

SUNFISH

Order Perciformes. Family Centrarchidae composed of freshwater basses and sunfishes. Native to North America, they inhabit freshwater lakes and rivers. Pumpkinseed (*Lepomis gibbosus*) is from 10–22 cm (4–8.8 in) long, and has been introduced into most regions of Europe. Diet consists of invertebrates and small fishes. In North America it is valued as both a food and sport fish. Green Sunfish (*L. cyanellus*) is from 10–30 cm (4–12 in) long. It has been introduced into Italy. Redbreast Sunfish (*L. auritus*) is from 12–24 cm (4.8–9.6 in) long and was introduced into West Germany. Bluegill (*L. macrochirus*) is from 15–23 cm (6–9 in) long. Banded Pygmy Sunfish (*Elassoma zonatum*) and Everglades Pygmy Sunfish (*E. evergladei*) are less than 5 cm (2 in).

The sunfishes, family Centrarchidae, are native to the United States and Canada but some species, including the Pumpkinseed (illustrated here), Green Sunfish and Redbreast Sunfish have been introduced into Europe. Members of this group prefer slow-flowing rivers and weedy lakes. The larger species (i.e. the Bluegill) are favourites with anglers while the smaller, brightly coloured species (i.e. the Banded Pygmy and Everglades Pygmy Sunfishes) are popular aquarium fishes. All sunfishes are distinguished by an undivided dorsal fin, the front portion of which is spiny, and three to nine spines in the anal fin. During the breeding season the males and females of some species take on a more intense colouring and engage in elaborate courtship rituals. The male makes a nest in sand or gravel and guards and tends both the eggs and the fry.

TARGETFISH

Targetfishes belong to the family Zeidae which also includes the well-known European and American John Dory. Dories are generally found in the middle regions of the open sea. Targetfishes live in the Red Sea and the Indian and Pacific Oceans but sometimes are also found in shallow coastal waters and among coral reefs. These fishes often associate with the jellyfish, especially when young, and take shelter from predators underneath its tentacles. They also keep together in shoals as a further means of protection. Their bright colour would seem to invite attention from carnivorous fishes, but the bold black stripes may help to break up the outline of their body, as in the vivid patterning of most coral reef fishes. The width and depth of colour of the stripes appear to vary with the age of the fish, becoming thinner and duller with age.

Order Zeiformes. Family Zeidae, the dories. Targetfish (*Zeus japonicus*) grows up to 90 cm (36 in) long. Found in the Red Sea and Indian and Pacific Oceans, it inhabits the middle regions of the sea and sometimes coastal waters and coral reefs. As in other dories there are no teeth, and along the lateral line there is a spiny shield. Diet consists of small aquatic animals and, in larger specimens, other fishes. A similar species occurs in waters of northern Australia. European John Dory (*Zeus faber*) grows up to 90 cm (36 in) long, and is found in the Mediterranean Sea and the Atlantic Ocean from the British Isles south to South Africa. American John Dory (*Zenopis ocellata*) is found along the eastern coast of North America.

ARCHERFISH

The Archerfish is native to South-east Asia and northern Australia. Because of its curious method of gathering prey it is often kept in aquaria. It has a groove in its palate which, when the tongue is pressed forcibly against it, projects a jet of water droplets from its mouth. This fish captures its food (mainly insects that dwell above the surface of the water) by poking its long snout out of the water, aiming a stream of water at an insect and knocking it out of the air. Archerfishes have a deep, squat body with both the dorsal and anal fins set well back towards the caudal fin. The most common species and popular aquarium species (*Toxotes jaculatrix*, illustrated here) has a white body with dark bands running vertically from the back down the sides towards the belly.

Order Perciformes. Family Toxotidae, composed of 4 species all found in South-east Asia and northern Australia. They inhabit brackish waters, mangrove swamps and occasionally freshwater regions such as rivers. They are 15–30 cm (6–12 in) long. Diet consists of insects as well as spiders. The Archerfishes are capable of capturing prey over a distance as great as 1 metre (40 in). Archerfish (*Toxotes jaculatrix*) is about 15 cm (6 in) long.

ROCK BEAUTY

The Rock Beauty is one of the many species of angelfish which are so popular with aquarists. They are found in warm waters throughout the world and generally inhabit coral reefs. This fish occurs in the Caribbean Sea and in the tropical waters of the eastern Atlantic Ocean. It is usually solitary but may swim about in pairs. Each fish maintains a well-defined territory which it fiercely defends against intruders by emitting a variety of growls and other warning sounds. As in most other species the colouring of the young differs to some extent from that of the parents. Young Rock Beauties are almost completely yellow with a small black mark ringed with blue on their backs. The black area increases in size with growth, so that adults have a black body with a yellow head, tail and underparts. The fry often associate with other fishes, feeding on the external parasites on their bodies.

Order Perciformes. Family Chaetodontidae, composed of butterflyfishes and angelfishes, all with flattened, deep bodies and colourful patterns over the body. Rock Beauty (*Holacanthus ciliaris*) is 30–60 cm (12–24 in) long in the wild, but only about 15 cm (6 in) long in aquaria. Found in the Caribbean Sea and in tropical regions of the eastern Atlantic Ocean, it inhabits coral reefs. Diet consists of sponges, algae, small crustaceans and sea anemones. Like most other angelfish species, the Rock Beauty has a spine at the end of the gill cover.

IMPERIAL ANGELFISH

Several different forms of the Imperial Angelfish exist in most of the world's tropical seas, particularly in the Indian and Pacific Oceans. They vary in both size and colour, ranging from reddish-purple or red with bold black markings to blue with white stripes arranged in semicircles. In several respects their habits are similar to those of the Rock Beauty (page 283); they are usually solitary or move about in twos and they defend their territory against other fishes. Their habitats, however, extend from coral reefs to rocky coasts which have an abundance of algae. In all regions they dwell fairly close to the bottom, from where they take their food. The diet includes algae (some of these fishes restrict their diet solely to this food), sea anemones, small crustaceans and molluscs.

Order Perciformes. Family Chaetodontidae. Imperial Angelfish (*Pomacanthus imperator*) ranges from 7–60cm (2.8–24in) long in its natural environment, while aquarium specimens usually attain a length of only about 25cm (10in). Found in most of the tropical seas but primarily in the Indian and Pacific Oceans, it inhabits coral reefs and rocky coasts where it often secretes itself among the rocks. As in the Rock Beauty and other angelfishes, the fry have a different appearance from that of their parents, so much so in this species that at one time they were considered to be a separate species. Other species of the genus *Pomacanthus* include the Blue Angelfish (*P. semicirculatus*) found in the Indian and Pacific Oceans, and the French Angelfish (*P. paru*).

Order Gasterosteiformes. Family Gasterosteidae. Stickle-backs are native only to the waters of the northern hemi-sphere and are found in North America, Europe and Asia. All have short spines in the first dorsal fin and many also pos-sess spines in the pelvic and anal fins. Three-spined Stickle-back (*Gasterosteus aculeatus*), from 4–11cm (1.6 4 4in) long, is found in many parts of Europe, Asia and North America. Colour-ing is normally a dull greyish-green but during the breeding season the male becomes a metallic blue-green with a bright red belly. Female grows plump and takes on a silvery sheen. She lays about 400 eggs; the fry are tended by the male for sev-eral days. Diet consists of in-vertebrates and fish fry.

THREE-SPINED STICKLEBACK

The Three-spined Stickleback is a popular and fascinating aquarium fish for it has an interesting and complex reproductive cycle. At the beginning of the breeding season the male marks out his territory among the vegetation and makes a shallow depression in the sand in which he builds a tubular nest of plant materials held together by a sticky secretion. Having selected a female, he performs a zig-zag dance and drives her towards the nest, biting at her fins and chasing her if she deviates from her course. After she has laid her eggs she is driven away and other females may be induced to lay in the same nest. The eggs take about seven days to hatch. During this period they are carefully guarded by the male, who frequently fans the water with his fins to aerate the incubating eggs. In its natural environment, the Three-spined Stickleback is found in a wide variety of habitats, from freshwater to shallow sea.

SCORPIONFISHES

As their name implies, the scorpionfishes are capable of inflicting an extremely painful and sometimes fatal sting with their sharp, pointed spines. Although some species are found in the Atlantic Ocean off the coast of North and South America, and in the Caribbean Sea, the most dangerous and venomous occur in the Indian and Pacific Oceans. One group, the zebrafishes, lurks among rocks and coral reefs. These fishes have beautifully striped bodies which camouflage them from their prey. They have separate, elongated rays on their pectoral and dorsal fins, the latter having venom glands at their base. After the fish has stabbed its prey, the poison flows along grooves in the spines and into the wound, paralysing small victims and causing considerable discomfort to humans.

Order Scorpaeniformes. Family Scorpaenidae, composed of about 300 species of scorpionfish and rockfish. All are marine fishes that usually dwell in deep water, although some are found in shallow coastal waters. Zebrafishes (*Pterois spp*) grow up to 37cm (14½in) long. Found in the Indian and Pacific Oceans and in the Red Sea, they inhabit coral reefs and rocky regions. Their colour is usually reddish-brown with white stripes. Close North American relatives include lionfishes (*Scorpaena spp*) which occur in warm waters of the Atlantic Ocean, and the California Scorpionfish (*S. guttata*) found in the warm waters of the eastern Pacific Ocean.

Order Pleuronectiformes. Family Pleuronectidae – dextral or right-eyed flounders. European Plaice (*Pleuronectes platessa*) grows up to 105cm (42in) long, with a weight of up to 9kg (20lb). As in other flatfishes, growth is dependent on a number of environmental factors. Found in the North Atlantic, especially in the North Sea, along the coasts of North Africa, and from Portugal to Norway, it is also present in Icelandic waters. The female may carry up to half a million eggs. Young adults live in shallow inshore waters, older adults occur to depths of 200 metres (656ft). Diet consists of worms, crustaceans, bristleworms etc. American Plaice (*Hippoglossoides platessoides*) grows up to 60cm (24in) long, and weighs up to 1.8kg (4lb).

PLAICE

The Plaice belongs to the order Pleuronectiformes which is composed of the four families of flatfishes. These are bottom-dwelling marine fishes with asymmetrical, laterally compressed bodies. Their eyes are on either the right or left side of the head, depending on the species. The flatfish larva resembles that of other fishes (it swims in an upright position and has a symmetrical body) and lives near the surface of the sea but when it has grown to a certain length it undergoes a complex metamorphosis during which it moves down to the sea floor. The Plaice larva inclines its body to the left; gradually the left eye moves or migrates towards the right eye, and the dorsal and ventral fins lengthen. When metamorphosis is complete, the Plaice measures about 14 millimetres (about half an inch) long; the right side of the body is uppermost and the underside has lost its pigmentation. European Plaice has a green body with orange spots.

REMORAS

Remoras are marine fishes found throughout the world in subtropical and tropical waters. These long, slim fishes have a flat, ridged suction disc on the head with which they attach themselves to other marine animals and objects such as boats. The disc, which is a modified dorsal fin, creates a vacuum between the fish and the host. By moving backwards the remora increases the suction; to disengage itself it moves forwards. Remoras feed on zooplankton, small invertebrates, and the food remains of their host, for whom they provide a service by ridding them of external parasites. Nine known species exist, the largest of which is the Sharksucker, a remora that associates with sharks, turtles and groupers. In certain regions of the world remoras have been used to capture large sea turtles. Attached by its tail to a rope, the remora would be lowered from a boat and fasten on to the underside of the turtle's shell.

Order Perciformes. Family Echeneidae. Remoras have an elongated body covered with tiny scales. The mouth is large and the lower jaw protrudes forwards. The teeth, which are small and sharp, are in bands on the tongue, gums and jaws. There is no swim bladder. The number of ridges on the suction disc varies according to the species. Remora (*Remora remora*) grows up to 38 cm (15 in) long with 18 ridges. It has a dark colouring with a white stripe down each side. It associates with sharks and other large fishes, and is found in most of the warm waters of the world. Sharksucker (*Echeneis naucrates*) can grow up to 1 metre (40 in) but averages 45 cm (18 in) with 21–28 ridges. It has a similar distribution to the Remora.

INSECTS

Insects are a class of invertebrates belonging to the Phylum Arthropoda (see page 327). With some 700,000 named species, the Class Insecta is the largest in the animal kingdom. It is divided into two subclasses, the Apterygota and the Pterygota. In the former group are placed the small and primitive wingless species such as silverfish. In these the young are very similar to the adults and undergo very little transformation as they develop. The second and larger group contains the more familiar winged species, although in some the wings are reduced or absent – these species lost their wings as they evolved. They therefore differ from the apterygotes which have never possessed wings. The Pterygota are further distinguished by having distinct larval and adult phases, the young undergoing a transformation known as metamorphosis between the first phase and the second.

The insect's body is divided into three sections: the head, thorax and abdomen. The head is the site of the feeding mechanisms (the mouth parts), the senses of touch and smell (the antennae) and sight (the eyes). The eyes may be compound, as in flies, with one eye on either side of the head, each composed of many facets or units, or simple, in which case there are generally three eyes (ocelli) placed near the back of the head. The mouth parts vary depending on the type of food and method of feeding but basically are composed of three parts: a pair of mandibles or upper, toothed jaws, a second pair of jaws, the maxillae, situated behind the mandibles, and a lower pair fused to form the labium or lower lip. Above the mandibles lies the labrum or upper lip. The second body segment, the thorax, consists of the prothorax, the section nearest the head, the mid-section or mesothorax and the hind section or metathorax. Each section has one pair of legs and in most winged insects the mesothorax and metathorax each possesses a pair of wings, membranous structures supported by a network of veins. The third and last segment, the abdomen, contains the digestive and excretory organs, and usually a pair of cerci which are thin, tail-like filaments that act as secondary antennae. In addition, the abdomen of females carries an ovipositor, the primary function of which is to deposit eggs, although in stinging insects it has become a stinging mechanism. The body, including the legs, is covered by a hard skin composed mainly of chitin, which is impermeable to water. This is in fact the exoskeleton or cuticle, found in all arthropods, that protects the internal organs and supports the body. It is also present in thin layers in the throat, breathing tubes and gut.

In insects, fertilization of the eggs takes place internally, the males possessing a penis for this purpose so that true copulation occurs, and the female deposits her eggs through the ovipositor. The larva hatches by breaking open the shell with a temporary structure called an egg-burster, or by merely lifting a lid or cap. As they grow and develop, insect larvae shed their outer skin to allow for their increased size; the number of times moulting takes place depends on the species. In insects with incomplete or direct metamorphosis, such as grasshoppers, the larva resembles the adult and is called a nymph. Insects such as beetles, butterflies and moths metamorphose in a more complicated way. As larvae they bear no resemblance to the adult, or imago, and as they grow and moult their appearance remains the same. When fully grown they change into a pupa encased in a hard cocoon. Outwardly, they appear to be inactive. Inside, however, dramatic changes take place in which the tissues and organs take on the form of the adult.

Insects, however, are among the most successful of creatures. Being both adaptable and usually inconspicuous, they have managed to survive, diversify and spread into every conceivable kind of niche.

MANTIDS

Order Dictyoptera, cockroaches and mantids. Suborder Mantodea, composed of over 1500 species of mantid. Mantids range in size from 1–16 cm (0.4–6.4 in) long. Praying Mantis (*Mantis religiosa*) is up to 7.5 cm (3 in) long. It is found in Mediterranean countries, parts of Asia and Africa, and in north-eastern North America where it has been introduced. Colouring is green, camouflaging it among foliage. The eggs of mantids are surrounded by a protective case known as the ootheca; it is made from a thick liquid secreted from the female's abdomen and is attached to stones, bushes and other objects. The nymphs may undergo as many as 12 moults before they become adults. A few females have been known to devour the males after mating.

Mantids most frequently occur in subtropical and tropical parts of the world. These species are usually beautifully coloured, particularly the flower mantids that in colour and shape resemble the flowers on which they live. All mantids are predatory insects with powerful forelegs adapted for capturing and grasping prey. They have extremely mobile heads and strong, biting mouth parts. Small species generally restrict their diet to insects but some of the larger ones prey on small lizards, frogs and young birds. Only a few species occur in Europe, the most common being the Praying Mantis of the Mediterranean region. Like other mantids, it does not pursue its prey but remains remarkably still, concentrating on its surroundings and awaiting the approach of an insect which it then seizes with lightning speed. This insect is named for the manner in which it holds up its forelimbs when at rest, as if in prayer.

291

STICK INSECTS

Order Phasmida, comprising some 2000 species of leaf and stick insects. Most are found in Asia. Some species (the leaf insects especially), so perfectly resemble the plant on which they live that they are mistakenly bitten by other insects. The manner in which these insects imitate plant species is known as cryptic mimicry. A few species are also able to change colour, being paler in the day and becoming darker at night. Mediterranean Stick Insect (*Bacillus rossii*), about 85mm (3.4in) long, is mainly nocturnal. A diet of plant material is common to all members of this order. Laboratory Stick Insect (*Carausius morosus*). North American species include the Walking Stick (*Diapheromera femorata*).

Stick insects have a slender, very elongated body and are remarkable for the manner in which they resemble twigs and stems, keeping so still that they may frequently be mistaken for part of the plant. They are also one of the few species of insect that are able to regenerate lost or damaged limbs, replacing them at the next moult. Most stick insects are found in subtropical and tropical regions of the world, although one species occurs in southern Europe. The species commonly kept as a pet and also used for scientific research is the Laboratory Stick Insect, a native of India. As in some other species, the females can reproduce parthenogenetically, producing young – always females – from unfertilized eggs. The eggs are hard and usually resemble seeds; they can remain dormant for long periods. The young resemble the adults but are much more agile and active.

Order Orthoptera, grass hoppers, bush crickets and true crickets. (Cockroaches are sometimes assigned to this order; grasshoppers and their allies are sometimes placed together in the order Saltoria.) In North America bush crickets are known as katydids. Two species common in Europe are the Great Green Bush Cricket (*Tettigonia viridissima*), from 45–57mm (1.7–2.2in) long, and the Wartbiter (*Decticus verrucivorus*), 35mm (1.37in) long. The Great Green Bush Cricket is found in Europe, North Africa and much of Asia. As in other species, the eggs are laid during winter and the nymphs emerge in spring. A similar North American species is the Big Green Pine-tree Katydid (*Hubbellia marginifera*).

BUSH CRICKET

Bush crickets belong to the same group of insects as the grasshoppers, the Orthoptera. They are sometimes referred to as the short-horned grasshoppers and long-horned grasshoppers, the names referring to the length of the antennae. In the bush crickets, the antennae are longer than the combined length of the head and thorax, and are thinly filamentous. Grasshoppers have comparatively short antennae. Only male bush crickets and grasshoppers are able to sing or stridulate. Bush crickets produce their song by rubbing a hard scale on the right forewing (the tegmen) against a protruding nerve on the left tegmen. Both grasshoppers and bush crickets have very long, powerful hind legs adapted for jumping. The female bush cricket may be distinguished by the long, pointed ovipositor. Bush crickets have ears on the tibiae of the forelegs. In grasshoppers the ears are located on either side of the abdomen.

FIELD CRICKET

The Field Cricket is a true cricket of the group Gryllidae, found in grassy places such as the edges of fields. Like other crickets it has long tail filaments, or cerci, and long thread-like antennae. The males stridulate by rubbing the forewings back and forth against each other. Like other stridulating insects, they sing to attract the females during the breeding season. Field Crickets are terrestrial creatures and dig holes in the ground where they spend the day, emerging at night to feed on plants. The forewings lie flat against the body, but they are atrophied and are not used for flying. A close relative is the Mole Cricket which is also terrestrial and has front limbs adapted for digging; it flies clumsily just above the ground.

Order Orthoptera. Family Gryllidae. Field Cricket (*Gryllus campestris*), about 24mm (1in) long. Found in central and southern Europe, North Africa and western Asia. Inhabits edges of fields, meadows and grassy slopes. Both nymphs and adults feed mainly on plant material. Nymphs emerging as adults in the spring. Seven species of this genus, including the Wood Cricket, are found in North America. Mole Cricket (*Gryllotalpa gryllotalpa*), from 38–42mm (1.4–1.6in) long, is found in Europe, western Asia and North Africa. It inhabits well-aerated soil, often near water. Diet consists of insects, worms and roots. It is sometimes considered a pest as it destroys crops. Female lays 200–300 eggs in an underground chamber. The larval stage lasts for 2 years.

TERMITES

Order Isoptera, comprising six families found mainly in the tropical regions of the world. Termites have extremely well-developed systems of communication which are mainly dependent on chemical substances called pheromones, contained in the saliva and excrement. Messages are transmitted when saliva is exchanged and the droppings eaten. The queen is a large, worm-like egg-producing machine, laying up to 30,000 eggs in one day. At certain times she produces larvae that are winged and sexually productive. When conditions are right, usually at the start of the rainy season, these termites swarm out of the nest. Those that are not eaten by predators pair off and then construct a royal cell, the beginning of a new colony.

Termites have a very highly developed social system and live in huge colonies, the largest containing up to one million individuals. As in bees, true ants and wasps, the inhabitants are all the offspring of a single pair of adults (the queen and king, or primary reproducers) and are divided into different groups or castes, each one differentiated by anatomical and functional features. Both the workers and soldiers are wingless and sterile. The soldiers function only to guard the colony against invaders; the workers, who are blind, construct and repair the nest, feed all the other members and care for the queen, her eggs and the larvae. In addition there may be a group of smaller termites known as supplementary reproductives whose duty it is to take over the reproductive process should the fertile adults die.

MAYFLY

Mayflies may be seen by streams, rivers and lakes from May into the summer months, although individual mayflies live for only a short period (this may be several hours or, in some, several days). For this reason they are called the Ephemeroptera, from the Greek *ephemoros*, meaning living for a day. The sole function of the adults is to reproduce. They possess only vestigial mouth parts and so are unable to feed. They are characterized by long tail filaments, or cerci, at the end of the abdomen, very short antennae and wide, transparent forewings and much smaller hind wings. The nymphs are entirely aquatic and are equipped with gills on the sides of the abdomen. They are herbivorous and feed on algae and other plant materials. When fully grown, the nymph climbs from the water and moults, emerging as a subimago with fully developed wings but legs and cerci that are not fully developed until the final moult.

Order Ephemeroptera, comprising more than 1000 species found in all parts of the world. They inhabit clear, running water or sandy water depending on the species. Mayfly (*Ephemera danica*), up to 40mm (1.5in) long, is found in central Europe and Britain. It inhabits small, clear ponds. As in other species, the males swarm and perform a ritual dance in the air prior to mating, which takes place in flight. The female either scatters her eggs on the water or submerges and deposits them in the water. The nymphs of this species live for up to two years before transformation into the subimago stage, and usually burrow in the sandy or muddy bottom where they feed on organic particles.

DRAGONFLIES

Dragonflies compose the order Odonata which is divided into two main groups, the Anisoptera and the Zygoptera. (There is also a third suborder, the Anisozygoptera, consisting of only one genus.) To the Zygoptera belong those species commonly known as the damselflies and demoiselles; they are smaller than the Anisoptera and have very slender bodies. When at rest they hold their wings over their backs. Their larger and sturdier relatives rest with their wings held out to the sides and many fly with quick, darting movements. All adult dragonflies are carnivorous and catch their prey on the wing, grasping them firmly in their legs and devouring them with their very powerful, biting mouthparts. The nymphs are aquatic and also predacious. They are equipped with a special device known as a labium for capturing invertebrates and fish fry. The labium is folded under the head but can shoot out rapidly when needed.

Order Odonata, dragonflies and damselflies. Suborder Zygoptera includes the Demoiselle Agrion (*Agrion virgo*), length 58–63 mm (2.2–2.5 in); wing span 70 mm (2.7 in). Found in Europe, parts of Asia and North Africa. Inhabits fast-flowing streams and rivers lined with trees. Suborder Anisoptera includes the Common Aeshna (*Aeshna juncea*), up to 76 mm (3 in) long. It inhabits lakes and ponds in pine woods, and gardens. In flight, it both hovers and darts. Members of the Odonata are characterized by large eyes and have good vision. The mating process takes place in flight or when the pair are settled on a plant stem. The male grasps the female with his tail claspers, depositing the semen with a special copulatory organ which is filled with semen.

SHIELDBUGS

Order Hemiptera, comprising true bugs and their allies. Divided into two suborders: the Heteroptera, which includes the shieldbugs, and the Homoptera which includes the aphids. Shieldbugs belong to the family Pentatomidae, in most of which the antennae are divided into five segments. Green Shieldbug (*Palomena prasina*), 12–15mm (0.47–0.59in) long. Found in Europe, temperate parts of Asia, North America and North Africa. It favours deciduous trees such as hazel and birch. Female lays her eggs on leaves in batches; nymphs undergo several moults before becoming adults. Forest Bug (*Pentatoma rufipes*), 11–16mm (0.47–0.59in) long. Distinguished by a yellow spot at the point of the shield. Found especially on oak and fruit trees.

Shieldbugs get their name from the triangular-shaped shield on their dorsal surface. They are also known as stinkbugs in North America because of their offensive odour which serves as a warning device to their predators. Although the base of the forewings is leathery, the tips are membranous, distinguishing them from beetles in which the forewings are uniformly hardened. The mouth parts of shieldbugs and all other members of the Order Hemiptera have modified into a thin, elongated organ for piercing and sucking. These insects are predominantly plant-suckers which extract juices from fruit, leaves and other plant parts. Some, such as the Forest Bug illustrated here, attack insects and other animals. The Green Shieldbug or Stinkbug is common in Europe and North America and inhabits herbs, bushes and deciduous trees. Bright green in spring and summer, it becomes a bronzy-red in the autumn.

POND SKATER

Order Hemiptera. Suborder Heteroptera. Family Gerridae. Common Pond Skater (*Gerris lacustris*), about 12 mm (0.47 in) long. The female lays her eggs on aquatic plants, to which they adhere by means of a sticky coating. Water measurer (*Hydrometra stagnorum* and *spp*), 9–12 mm (0.35–0.47 in) long, spends part of its time on land near the water. It moves in both habitats with a slow walking movement. In this bug the legs are all about the same length and its forelegs are not modified for capturing prey. Diet consists of small aquatic insects and larvae which they pierce with their mouthparts. They usually mate on water but the eggs are attached to plants and rocks above the water line.

The order Hemiptera contains a number of aquatic and semi-aquatic bugs that live in or on the surface of water. Two common surface-dwellers are the Pond Skater or Water Strider and the Water Measurer or Gnat, both of which may be seen on ponds and lakes and other quiet or slow-flowing waters. The forelegs of the Pond Skater are relatively short and are used to grasp prey such as midges and other insects that fall on to the water. The much longer middle and hind legs provide mobility; the former are used in a rowing movement and the latter act as rudders. Their ability to move on the surface is owed to the surface tension of water, which creates a film capable of bearing the weight of the insect. A fine, smooth, hairy covering on the body and legs prevents water penetrating even when the insect becomes submerged. Some Pond Skaters are wingless but others can fly. They hibernate among plants and stones.

CICADA

The two preceding members of the Hemiptera, the Shieldbug and the Pond Skater, belong to the suborder Heteroptera. The insects of the second, larger suborder Homoptera are much more varied than their relatives and are contained in two groups: the Auchenor-rhyncha which includes the cicadas, and the Sternor-rhyncha which includes the aphids (page 302). The Homoptera, like all Hemiptera, possess sucking and piercing mouth parts, but their wings are either entirely membranous or entirely leathery, and the insects tend to be much smaller.

The family Cicadidae contains some 1500 species of cicada, most of which occur in warm-temperate and tropical parts of the world. Like grasshoppers, bush crickets and true crickets, the males sing loudly to attract the females, and of all the musical insects they have the loudest and most highly developed sound instruments. In the forepart of the abdomen there are two small drums or plates called tymbals which are attached to strong muscles. The muscles pull the tymbals and then release them so that they snap back into place with a loud click and then vibrate. Air cavities in the abdomen act as resonators and amplify the sound. The action of the muscles may take place from about 100 to 500 times in a second, producing an extremely rapid song. It is interesting to note that each species of cicada has a song that in rhythm and tone is unlike any other. Experts are able to distinguish each species by the noise it emits.

Adult cicadas live on plants, including trees, from which they suck the sap and for this reason are sometimes considered to be pests. They are generally black or brown and when disturbed often flatten their bodies against the branch or bark so that they are difficult to see. They also have keen sight and strong wings, and sometimes use flight to evade danger. The few species found in Europe include the Mountain Cicada which occurs in the more southerly parts of Europe and in southern Britain where it is known as the New Forest Cicada.

Order Hemiptera. Suborder Homoptera. Family Cicadidae, comprising some 1500 species found in the warm-temperate and tropical regions of the world. The female has a strong ovipositor and deposits her eggs in the woody parts of plants. Newly-hatched larvae fall off or move down into the soil, burrowing with their powerful forelegs. They feed on the roots of plants, extracting the juices, and like the adults can cause considerable damage. The development of the larvae is slow and in most species they take several years to emerge as adults. Larvae of the North American Periodical or Seventeen-Year Cicada (*Magicicada septendecim*) may take up to 17 years to reach the adult stage. This generally occurs in the northerly specimens, the ones in the south taking 13 years. The adults live for only a brief time, 4 to 6 weeks, and do not eat. The New Forest Cicada or Mountain Cicada (*Cicadetta montana*), 25 mm (1 in) long, prefers oak woods.

APHIDS

Order Hemiptera. Suborder Homoptera. Family Aphididae. Rose Aphid (*Macrosiphum rosae*), 2–3mm (0.07–0.11in) long. Bean Aphid (*Aphis fabae*) is of similar length. The reproductive process of aphids is extremely complex. In the autumn many aphids die but some mate first and produce eggs which over-winter. The aphids that hatch in the spring are wingless females that are ovoviviparous (eggs hatch within the body) and can reproduce without having been fertilized by a male (parthenogenesis). Their young usually reproduce in the same way, but some of this new generation will be winged and will fly to other plants. During the summer both winged and wingless generations are produced, but they are usually all females. In the autumn the males hatch.

There are over 2000 species of this very small insect. They are abundant in most parts of the world and cause considerable damage to plants. With their sucking proboscis they take in sap, weakening the plant they feed on and causing distortion of the leaves and stunted growth. One of the better-known species, especially to gardeners, is the Greenfly or Rose Aphid which is often seen in the summer on roses and other cultivated garden plants. Like many other species it secretes a sweet, sticky substance called honeydew which is attractive to other insects such as ants (see page 324). Honeydew is composed of sugars and other substances found in sap which the aphid's digestive system can make use of only in small amounts so that the excess passes out of the body. The Rose Aphid may be pinkish or green. Other species include the black Bean Aphid or Blackfly found on broadbeans and some other plants.

GROUND BEETLES

Beetles compose the order Coleoptera. With some 250,000 known species this is not only the largest order of insects but also the largest in the whole of the animal kingdom. The main characteristics of the group can be seen clearly in this illustration of the Violet Ground Beetle, particularly the hard forewings, or elytra, which cover and protect the hind body. These horny sheaths are not used in flight but are held out at an angle from the body, thus exposing the hind wings which are the actual organs of flight. The latter are tucked away under the elytra when they are not in use. In some species the hind wings are much reduced or absent, rendering the beetle flightless. The Violet Ground Beetle is a typical member of the family Carabidae; the long, slim legs are adapted for running quickly over the ground and the mandibles are sharply pointed and toothed. These are used for grasping and tearing prey.

Order Coleoptera, the beetles. Family Carabidae. Violet Ground Beetle (*Carabus violaceus*), 30–35mm (1.1–1.3in) long, is found in Europe. It inhabits gardens, hedges and woods, where it lives under stones and plant material. Like all ground beetles it is carnivorous in both the larval and adult stages, with mouthparts adapted for chewing and biting. Adaptations to a predatory life include the large eyes and long, sensitive antennae. The body is black with a metallic violet sheen. The Pupa-predator (*Calosoma sycophanta*) is an attractive species with metallic red and green elytra and a blue shield on the underside. It feeds on caterpillars and is found mainly in deciduous forests. It was introduced into the United States to control caterpillars.

RED AND BLACK BURYING BEETLE

Order Coleoptera. Family Silphidae, comprising about 2000 species of carrion beetle. Red and Black Burying Beetle (*Necrophorus vespilloides*), 15–20 mm (0.59–0.78 in) long, is found in Europe, including Britain. Once a dead animal has been buried and covered over, the female beetle masticates pieces of its flesh and shapes the chewed material into a ball. This is placed in an underground chamber in which she lays her eggs. The newly-hatched larvae feed on the masticated food and later make their way to the corpse on which they continue to feed. When they are fully developed they burrow deeper and pupate for the winter.

Like many other kinds of animal, the insects have their scavengers which play an important role in ridding the environment of dead animals and litter which might otherwise spread disease. Some of these insects occur in the large order Coleoptera, which includes the scavenging carrion beetles and the dung beetles (page 309). The Red and Black Burying Beetle or Sexton Beetle belongs to the former group and its whole life-cycle is dependent on the carrion of small animals. Several burying beetles working together can rapidly bury a corpse in the ground, usually digging the ground beneath it with their legs and using their heads to shovel the soil aside. If the ground proves too hard to excavate, they move the dead animal by turning on their backs underneath it, grasping it in their jaws and pushing it along to an area with softer soil.

LADYBIRDS

Order Coleoptera. Family Coccinellidae, the ladybirds (known as ladybugs in North America). Seven-spotted Ladybird (*Coccinella septempunctata*), 6–7mm (0.23–0.27 in) long, has red wing covers with 7 black spots. Female lays her small, yellow eggs on the underside of leaves that are infested with aphids, thus ensuring that the larvae are provided with an abundance of food. The yellow-spotted, blue larvae pupate after 3 weeks, attaching themselves to the underside of a leaf with a sticky secretion. Newly-emerged adults are at first soft-bodied. Two-spotted Ladybird (*Adalia bipunctata*), 4–5 mm (0.15–0.19 in) long, is usually red but may be black with 4 spots (*A. bipunctata quadrimaculata*) or black with 6 red spots (*A. bipunctata sexpustulata*).

These attractive and popular beetles are especially welcomed by gardeners, for both larvae and adults consume large quantities of aphids and scale-insects. Over 4000 species composing the family Coccinellidae are found in almost every part of the world. Their bright colours warn predators of their unpleasant taste, as does the fluid they secrete when handled. The fluid is in fact blood which automatically oozes from joints and other body parts when the insect is threatened. This characteristic, known as reflex bleeding, occurs in other insects such as the Bloody-nosed Beetle. The very common Seven-spotted Ladybird passes the winter in crevices in old walls or under bark and stones. Similar places are used by the Two-spotted Ladybird but in this species 50 or more individuals hibernate together, assembling in groups before they withdraw.

LONGHORN BEETLES

The large group of longhorn, or longicorn, beetles are wood-boring insects that may cause considerable damage to trees, particularly broadleaved trees. The adults are distinguished by very long antennae, sometimes longer than the combined length of the head and body, and many of them are attractively coloured and patterned. They are often found on or under the bark of trees, where they live in the larval stage, or on leaves and nearby flowers. It is the larvae that bore into branches and tree trunks, excavating long tunnels in the wood. Generally, each species restricts itself to certain trees such as willow (Musk Beetle) and poplar (Large Poplar Beetle, illustrated here). Longhorn beetles are found throughout the world in temperate, subtropical and tropical regions. They are most abundant in the last two areas, in which occur the largest and some of the most splendid specimens.

Order Coleoptera. Family Cerambycidae, comprising some 1800 species of longhorn or longicorn beetles. European species include the Large Poplar Longhorn (*Saperda populnea*), 18–28 mm (0.7–1.1 in) long, and the Musk Beetle (*Aromia moschata*), 20–32 mm (0.78–1.2 in) long. This beetle is greenish and is named for the pleasant odour it emits. The adults may squeak when handled. Another European species, the Wasp Beetle (*Clytus arietis*), is found in the larval stage in posts and palings of fences. The adults mimic the black and yellow colours of the wasp; this aids in deterring predators. Many longhorn larvae are eaten by woodpeckers. The largest species *Titanus giganteus* occurs in northern South America and measures 15 cm (6 in).

COLORADO POTATO BEETLE

Order Coleoptera. Family Chrysomelidae comprising some 25,000 species of leaf beetle found in most parts of the world. All have highly arched bodies and many exhibit extremely beautiful, metallic colours that give them a jewel-like appearance. Colorado Potato Beetle (*Leptinotarsa decemlineata*), 10–12 mm (0.39–0.47 in) long. Native to North America, it is now also found in most parts of Europe. The female lays between 500 and 800 eggs on leaves. From these emerge the soft-bodied, bright red or orange, black-spotted larvae which are voracious feeders. The orange-coloured pupae develop underground and emerge as adults in the autumn. Red Poplar Leaf Beetle (*Chrysomela populi*) is found on poplar and sallow.

This beetle causes devastation and destruction to potato crops, both the larvae and adults feeding on the foliage, and if seen its presence should be reported immediately. Prior to the 19th century it occurred only in western North America where it lived and fed on species of the nightshade family. With the colonization of the continent by Europeans and the introduction by them of the potato plant (a member of the nightshade family), the beetle soon spread to other areas. In 1922 it was accidently introduced into France from where it spread to many parts of Europe and, in 1933, to England. Some other members of this family, the Chrysomelidae or leafbeetles are also injurious to plants. They include the Asparagus Beetle and the Red Poplar Leaf Beetle.

STAG BEETLE

Some 800 species of stag beetle are found in most parts of the world. In these insects, especially the males, the mandibles (upper mouthparts) have developed into long, toothed devices that resemble the antlers of a stag. Stag beetles live on wounded and decaying trees, the adults feeding on sap and the larvae on the wood. The size of the adult male varies from one individual to another. In battles with rival males it is usually the strongest and largest who wins, thus ensuring that only the fittest specimens will reproduce. The females are smaller and darker than the males; they also have less obvious antlers. They lay their eggs in rotting wood on which the larvae live and feed for three, sometimes as many as five, years before pupating. Stag beetles belong to the large superfamily of beetles, the Lamellicornia, named for the manner in which the tips of the antennae have become flattened into plates, or lamellae.

Order Coleoptera. Superfamily Lamellicornia. Family Lucanidae, the stag beetles. European species include the Common Stag Beetle (*Lucanus cervus*), between 25–80 mm (0.98–3.1 in) long. It is found in Asia and Europe, including Britain where it is the largest beetle. It inhabits oak and sometimes beech. Because it lives on the wood and sap of trees, it is sometimes considered a pest although in fact it does not attack healthy specimens. A close North American relative, *L. elephas* of the southern United States, has similar habits and appearance. One of the largest stag beetles is the Giraffe Stag Beetle (*Cladognathus giraffa*), up to 90 mm (3.5 in) long. It is found in India and Java.

DUNG BEETLES

As the Red and Black Burying Beetle (page 304) is dependent on carrion, so the dung beetles of the family Scarabaeidae are dependent on dung. These rather attractive, rounded little insects are found near the droppings of sheep, cows and horses. Species of the genera *Geotrupes* and *Aphodius* are common in Europe. Included in the former group is the Dor Beetle, sometimes locally named the Lousy Watchman, for it is much infested with tiny mites. The name 'dor' comes from an Old English word meaning to drone or hum; the Dor Beetle flies at night, humming as it goes to a dung heap. The female burrows under the dung, rolling it into balls or plugs which she places in an underground chamber. When a store of balls has accumulated she lays an egg in each one. In some species several beetles work together to roll a ball of dung over the ground to the burrow.

Order Coleoptera. Family Scarabaeidae comprising more than 20,000 species which occur in every part of the world except the Arctic and Antarctica. Many of the dung beetles are beautifully coloured with metallic blues, greens, violets and reds. Dor Beetle (*Geotrupes stercorarius*) is 16–24 mm (0.62–0.94 in) long. The Dung Beetle (*Aphodius rufipes*), 11–13 mm (0.43–0.51 in) long, and other members of this family lay their eggs in dung. The dung beetles are able to locate dung with their acute sense of smell. In the United States they are known as Tumblebugs. The Sacred Scarab Beetle was held in high regard by the ancient Egyptians.

CHAFERS

Like dung beetles, the chafers belong to the family Scarabaeidae. They include such well-known species as the Cockchafer or Maybug illustrated here, and the Rose Chafer. Cockchafers live in trees and shrubs. They may often be seen on early summer evenings flying in groups around tree tops or towards lighted windows, the males making a humming or buzzing sound as they fly. They feed on leaves with their strong, biting mouthparts. The females deposit their eggs in the soil in batches of 10 to 15 and the larvae hatch after three weeks. Known to farmers and gardeners as the white grub, the larva is whitish with a brown head and is usually crescent-shaped; the three pairs of legs are situated far forwards on the body. Cockchafer larvae are particularly fond of the roots of potatoes, wheat and other grasses. When the larvae are numerous they cause considerable damage to crops and grasslands. The adults too may destroy enormous quantities of leaves, stripping trees and bushes of their foliage.

The European Rose Chafer is an attractive metallic green on the upper side and coppery-red beneath; the elytra are dotted with white markings. This chafer may be found feeding on flower heads particularly roses, on bright summer days. The larvae of this species feed on rotting tree stumps and roots. Like Cockchafer larvae, they take several years to become fully developed. Close relatives in other parts of the world include the rose beetles of Australia, one of which has glossy brown elytra with yellow markings in a symmetrical design, and the splendid rose or flower beetles of the tropics which are among the most beautiful of all insects. Some of them have colours and patterns comparable to those of butterflies and moths. They include the large Goliath Beetles of Africa, which are much valued by collectors. Also prized by collectors is another member of the Scarabaeidae, the Hercules Beetle of Central America. In this beetle the thorax and head are elongated to form two long horns, one over the other. It is the largest known species of beetle; the males grow to a length of 17 centimetres (6 inches), including the horns.

Order Coleoptera. Family Scarabaeidae, comprising some 20,000 species. Cockchafer (*Melolontha melolontha*), 25–30 mm (0.98–1.1 in) long, is found in Europe and Asia. Like stag beetles, they have lamellae at the tips of the antennae. The larvae live for up to three years in the soil, emerging in the late summer as adults after pupating in a hole in the ground. Here they remain for the winter. Rose Chafer (*Cetonia aurata*) is 14–20 mm (0.55–0.78 in) long. The Australian Rose Beetle (*Eupecilia australasiae*) is of similar length. The Hercules Beetle (*Dynastes herculeus*) is up to 17 cm (6 in) long. The female lacks horns. It is found in tropical forests of Central America. North American June Bugs are close relatives of the Cockchafer and belong to the subfamily Melolonthinae.

SWALLOWTAIL BUTTERFLIES

The swallowtails compose a large and impressive family, the Papilionidae, the members of which are distinguished by and named for the 'tails' at the end of the hind wings. The Common Swallowtail is found throughout the continent of Europe but in Great Britain occurs only on the Norfolk Broads. This species is found on umbelliferous plants such as milk parsley and wild carrot, and may be seen in the summer sunning itself on warm stones. In the northern parts of its range it breeds once a year but in more southerly regions there may be two or three broods. The female lays her eggs on the underparts of leaves and the caterpillars pupate in the autumn, overwintering in the pupal stage. The Common Swallowtail may be found in flowering meadows and in woods, frequently near water.

Order Lepidoptera, comprising some 100,000 known species of butterflies and moths. (Butterflies may be distinguished from moths by their antennae which are club-shaped at the tips.) Family Papilionidae. Common Swallowtail (*Papilio machaon*) has a wing span of between 77 and 90mm (3–3.5in). It is found in continental Europe, the neighbouring parts of Asia, and in Norfolk, Great Britain. Other similar species of the genus *Papilio* are found in the Americas, India and Australia. Only a small number of this order have biting mouthparts. In the majority the maxillae have been modified to form a sucking tube, the proboscis, which is coiled up when not in use. There are four wings which are covered with overlapping scales.

Order Lepidoptera. Family Pieridae, comprising 1500 species. Green-veined White Butterfly (*Pieris napi*) has a wing span of 47–50 mm (1.8–1.9 in). It is found in most parts of the northern hemisphere, in Europe (including the British Isles), western and central Asia, North Africa and North America. Eggs, brownish and seed-like, are laid on the undersides of white mustard and charlock leaves. The caterpillars feed on the leaves and seed pods of these plants, reaching full growth after six or seven weeks. Caterpillars are greenish or bluish with black warts and pale, lateral stripes. They pupate on old wood and fences. The second brood pupates over winter emerging as adults in the spring.

GREEN-VEINED WHITE BUTTERFLY

A close relative of the Brimstone and Large or Cabbage White (page 314), this butterfly is common in gardens, meadows and the edges of woodland. The colouring may be brownish or yellowish but is usually white with black powder scales on the upper surface of the wings and green-edged veins on the lower surface, especially on the hind wings. This characteristic veining distinguishes the Green-veined White from the Small White with which it may be easily confused, particularly when in flight. The female is larger than the male and the markings are more extensive and intensive. In this species there are two broods each year, the first appearing in May and June, the second in July and August. The life-cycle of the Green-veined White is similar to that of the Large and Small Whites.

BRIMSTONE BUTTERFLY

The Brimstone or Sulphur Butterfly belongs to the family of whites, the Pieridae, which also includes the very common Large White or Cabbage Butterfly and the Clouded Yellow. The Brimstone occurs in the temperate parts of Europe and Asia, including the British Isles, and in parts of North Africa, inhabiting bushes in woodland and hedges. The eggs are laid on the underside of buckthorn leaves on which the larvae feed when they hatch in early summer. Males have bright, sulphur yellow wings, those of the females are paler and greenish; each wing ends in a small point and has an orange spot in the centre. These butterflies overwinter in the adult stage and in many regions are among the first species to emerge in the spring, often as early as February and March when the weather is mild.

Order Lepidoptera. Family Pieridae, which contains some 1500 species. Brimstone (*Gonopteryx rhamni*) has a wing span of 58–60 mm (2.2–2.3 in). Adults suck nectar from thistles and other flowers. Eggs are greenish. The caterpillar pupates in a chrysalis that resembles a curled leaf and is attached to a stem by a silken thread. Adults of the new brood appear in July and August. Large White or Cabbage (*Pieris brassicae*) has a wing span of 62–64 mm (2.4–2.5 in) and is found in Europe, Asia and North Africa. The caterpillars are usually found on cabbage, cauliflower and nasturtium and are considered pests. Close relative, the Small White (*P. rapae*), has similar habits and distribution. It has been introduced into North America and Australia.

MEADOW BROWN BUTTERFLY

The Meadow Brown belongs to the family Satyridae, a group of butterflies found throughout most of the world. Most are light or dark brown in colour and have eye-spots on their wings. The Meadow Brown inhabits grassy places such as meadows, fields and waste ground. From May to September the adults may be found on the flower heads of thistles, bramble and knapweed. The female deposits her eggs on grass blades; they are tiny, white and spherical. The caterpillar is green with a dark stripe down the back and lighter stripes on the sides. It generally feeds at night, becoming fully developed in May when it spins on a grass blade, emerging as an adult in June and July. In this species there is only one brood a year.

Order Lepidoptera. Family Satyridae, comprising some 2000 species, which also includes the heaths and graylings. This family is represented in North America by the wood-nymphs and in Australia by the bush-browns and others. Meadow Brown (*Maniola jurtina*) has a wing span of 50–55 mm (1.9–2.1 in). The female is larger than the male and more brightly coloured. It is found in Europe east to the Ural Mountains and in parts of North Africa and the Middle East. Because of its inconspicuous colouring it is often overlooked. In Britain it is one of the most common butter-flies inhabiting grassland.

315

EMPEROR MOTH

Order Lepidoptera. Family Saturniidae (Attacidae), the emperor moths, found in most countries except New Zealand. European Emperor (*Saturnia pavonia*). The female is much larger than the male, with a wing span of 70mm (2.75in); male wing span is 55mm (2.12in). It is found in continental Europe and Britain, and inhabits moors and heaths. In this species the adults lack functional mouthparts and therefore cannot feed. The males are distinguished by feathered antennae which enable them to detect unmated females over a distance of several kilometres. These virgin females possess scent glands at the rear of their body from which an odour is diffused into the air and detected by the males. The antennae of the females are smaller and thread-like.

Emperor moths, with some 1200 known species, are distributed throughout most parts of the world. They are medium to large moths with a conspicuous eye-spot on each of the four wings. In the European Emperor the forewings of the males are brownish and the hind wings are yellowish-orange. The female is a duller, uniform grey-brown. The eggs are generally laid on the leaves of sloe or bramble which provide food for the developing larvae. Black at first, the larvae become green with bright yellow or reddish hairy warts. When fully developed, the caterpillars pupate and overwinter in the pupal stage. The pear-shaped cocoon has an opening at the narrow end which is surrounded on the outside by stiff bristles, preventing predators from entering. Less widely distributed is the Greater Emperor Moth which in Europe occurs only in the south; it is the largest of the European Lepidoptera.

MOSQUITO

Mosquitoes belong to the order Diptera which also includes the true flies. With 80,000 known species it is one of the largest group of insects, many of which transmit diseases to man and other animals. The mosquito is well known for inflicting irritating bites but it is not widely known that it is only the females that attack. In fact, they do not bite but suck blood from their host. In many, but not all mosquitoes, a blood meal is necessary for the development of the eggs. In the Common Mosquito of Europe, the female lays her eggs in clusters on the surface of water. The larvae are legless and live just under the surface film breathing through pores which break through the surface of the water. They feed by vibrating special hairs on the head which causes water currents to carry small organisms into the mouth. The pupae are also active and aquatic; they swim by moving the tail end of the body up and down.

Order Diptera, composed of true flies, mosquitoes, midges and their allies. Family Culicidae, mosquitoes. Characterized by a slender body and legs, an elongated proboscis and one pair of narrow, scale-covered wings. Females feed on blood, nectar and the juices of fruits. Males lack piercing mouthparts and suck sugary fluids from plants. Urban species overwinter in houses and farm buildings. Members of the genus *Anopheles* transmit malaria, taking in the malarial organisms with the blood of infected victims and subsequently infecting others. Common Mosquito (*Culex pipiens*) of Europe rarely attacks humans; it prefers birds.

HOUSE FLY

The ubiquitous House Fly is a great menace to man, spreading and carrying diseases such as cholera, typhoid and dysentery; in areas where standards of hygiene are low it is one of the main causes of infant mortality. The House Fly feeds on decaying organic matter such as household refuse and excrement. Its feet and mouthparts become infested with germs which are transmitted to human food by the visiting fly and so passed on to the consumer of the food. Under warm, favourable conditions the House Fly breeds rapidly, usually laying its eggs on horse manure or other dung and on rotting vegetable matter. In as little as three weeks the eggs hatch and produce a new brood of sexually active adults.

Order Diptera. Family Muscidae, which includes the House Fly and Tsetse Fly. Like other members of the Diptera the House Fly (*Musca domestica*) has one pair of functional wings which are membranous and transparent. The hind wings are modified into filament-like structures with a knob at the end; known as halteres, they are used to maintain equilibrium. The mouthparts are modified to take in liquid food only. In this species the larvae or maggots are legless, with a very reduced head. The tail end of the body is broad and possesses two pairs of breathing organs. When fully developed, the larva retains its final skin which hardens and darkens into a capsule called a puparium. The larva pupates inside this barrel-shaped structure.

ICHNEUMON WASP

Order Hymenoptera. Suborder Apocrita. Family Ichneumonidae, with some 10,000 species found mainly in temperate zones. Ichneumon Wasp (*Rhyssa persuasoria*) is up to 35mm (1.4in) long and has an ovipositor of similar length. Found in most parts of Europe and in North America, it was introduced into New Zealand after the Giant Wood Wasp (*Urocerus gigas*) was taken there accidently. Woodwasp larvae cause considerable damage to forests and woodland but are controlled to some extent by the Ichneumon Wasp. Other parasitic species include *Perithous divinator* which has a more fly-like appearance. Its larvae feed on the larvae of wasps that live on raspberries and hops.

Ichneumon wasps, sometimes incorrectly called ichneumon flies, are related to bees and ants but belong to the group known as the Parasitica; their larvae live as parasites on other insects. The species *Rhyssa persuasoria*, often called the Pipecleaner, is found in the coniferous forests of Europe. The female has an extremely long, needle-sharp ovipositor which is used for depositing her eggs. These are laid in the wood of coniferous trees that are already inhabited by woodwasp larvae. The ichneumon wasp probably locates the exact position of the woodwasp larvae by using her sensitive antennae. She drills through the bark and wood, often to a depth of 6 centimetres (2.4 inches). She then lays a single egg on each larva. When it hatches, the ichneumon grub feeds initially on the fatty tissues and minor organs of the woodwasp larva finally consuming it completely, when it has made its way to the surface.

BEES

Bees are divided into two groups, social bees such as the Bumblebee and Honey or Hive Bee, and solitary bees such as the leafcutters. The latter group contains the majority of the Apoidae. Solitary bees meet with the opposite sex only during the mating season and the females usually have no contact with the young which are enclosed in a cell and provided with bee bread (pollen mixed with honey). Their nests are made of vegetation (the leafcutters use carefully cut pieces of leaf) and built either in the ground or in a hollow stem. Social bees, as their name implies, live together in colonies which in the Honey Bee may hold up to 30,000 individuals. They build their nests from wax produced in scale-like formations by the workers. The colony consists of a single, egg-producing queen, workers (who are sterile) and drones (males). The Honey Bee is the only species to build a permanent nest.

Order Hymenoptera, comprising over 100,000 species of ants, bees and wasps. Suborder Apocrita. Superfamily Apoidae. The Honey Bee (*Apis mellifera*) is dark brown with a slightly hairy body and transparent wings tinged with brown. Four races are distinguished, two of which are found in the wild and two found only in artificial hives. Bumblebees or Humblebees (genera *Bombus* and *Psithyrus*) are somewhat larger and have a very furry body, usually striped dark brown and bright yellow. They are found in Asia, northern Africa and the Americas, and have been introduced into Australasia. At the beginning of winter all members of the colony die except the new fertilized queen who hibernates until spring when she founds a new colony.

WASPS

Wasps are found in every part of the world. Although they have a bad reputation, many species destroy serious pests and are therefore beneficial to man. Like bees, wasps are composed of social and solitary species. The most familiar are the species of true wasps belonging to the family Vespidae. They have highly developed social systems and include the Common Wasp, German Wasp and Hornet. These insects differ from bees in many aspects of their behaviour, particularly the construction of the nest and the feeding of the larvae. New colonies are established each year and there are never incidents of swarming. The only member of the old colony to survive winter is the new, fertilized queen who emerges in the spring and begins to gather material for the nest. This is constructed of a material known as wasp paper, made from wood by the queen. She cuts pieces of wood with her sharp mandibles. The wood is then mixed with saliva as she chews it. The nest may be built in the ground (e.g. the Common Wasp) or in hollow trees, rafters of buildings and banks (e.g. the Hornet).

The Common Wasp begins to build by papering the roof of a cavity and hanging a pillar from this disc-shaped structure. At the bottom of the pillar she constructs several cells to house the first larvae. The cells are open at the bottom and the single egg in each adheres to the top of the cell. The insects that result from the first laying are workers who take over the business of nest-building, caring for the queen and for the young. As the colony grows additional cells are added, so that when the nest is complete it is a complex system of hexagonal cells on many levels, each level supported by a paper shell. Throughout the summer only workers emerge from the eggs. The larvae are fed on other insects and their grubs which are first masticated by the workers. At the end of the summer special 'royal' cells are constructed and from these emerge female, fertile wasps that will be the queens in the following year. Males are also produced at this time from unfertilized eggs. As autumn sets in the colony begins to disintegrate, the males and females mate outside the nest and during the following weeks most of the adults die.

Order Hymenoptera. Suborder Apocrita, divided into two groups, the Parasitica (page 320) and the Aculeata, or stinging insects. In these insects the females have an ovipositor modified into a defensive organ which is not used to lay eggs. The stings of bees are barbed. When a bee withdraws from its victim, the sting remains in the flesh together with part of the abdomen so that the insect dies shortly after. Wasps, however, have a smooth sting; they can withdraw it and use it again. Common Wasp (*Vespula vulgaris*), 10–14mm (0.39–0.55in) long, is found on flowers and fruit. German Wasp (*V. germanica*), 15–20mm (0.59–0.78in) long. The Hornet (*Vespa crabro*) is much larger – 22–30mm (0.86–1.1in) but rarer. It is usually found only in well-wooded areas.

WOOD ANT

Order Hymenoptera. Suborder Apocrita. Family Formicidae comprising 3500 species of ants which are distributed throughout the world. Wood Ant (*Formica rufa*), 5–11mm (0.43–0.59in) long, inhabits coniferous woods, especially those with pine trees. The nests may be up to 1 metre (3.3 feet) high with a circumference of 8 metres (26 feet). Diet consists of insects and other small invertebrates, and honeydew gathered from aphids. These ants have sharp, biting mouthparts but their main means of defence is the formic acid secreted by glands situated at the hind end of the body. The acid immobilizes other insects and small animals, and can cause discomfort of the skin in humans.

Ants form the third large group of social insects of the order Hymenoptera but, unlike bees and wasps, there are no solitary species, all of them forming highly developed societies. The Wood Ants are a particularly interesting group. They are prevalent in coniferous woodland where they construct a mound or nest using conifer needles, sticks and twigs. The living and breeding quarters are situated underground and consist of innumerable tunnels and chambers; the ants gain access through gateways distributed around the mound and at night these are closed off and guarded by sentinels. Neatly kept, straight 'roads' run from the nest, some to nearby trees which are infested with aphids; these are used by the ants as a source of honeydew. Although there may be several nests in the same vicinity, the members of each colony stay within their own territory and do not appear to compete with each other.

INVERTEBRATES

The invertebrates (animals without back-bones) are a large and extremely hetero-geneous group of animals that contains the majority of living species; these comprise approximately 27 smaller groups or phyla. Although invertebrates are often termed the lower animals, they cannot be considered unsophisticated for many of them have very complex structures.

Of the numerous phyla, six of the major ones are represented in the following pages: the echinoderms, annelids, arthropods, mol-luscs, coelenterates and protozoans. As it is impossible to mention every existing group, emphasis is placed here on describing those phyla to which the species described in the main text belong. The major class of ar-thropods, the insects, is discussed separately on pages 289 to 324.

Protozoans, Sponges and Coelenterates

Life began in the sea. The first animals are thought to have emerged there about 2000 million years ago, possibly developing from simple plant organisms that in turn had evolved from primitive bacteria and viruses, the earliest life-forms. These 'first animals', which is the meaning of the word protozoa, are single-celled structures in which all the activities necessary for maintaining life are carried out in a tiny, usually microscopic, unit. It is interesting to note that the study of protozoans has been greatly helped by the recent development of the electron micro-scope which enables scientists to examine them in greater detail. In some protozoa the demarc-ation between plant and animal is slight. This has led some scientists to place all unicellular and simpler organisms in a third kingdom, the Protista.

Although cells are often thought of as simple structures with a nucleus surrounded by cytoplasm, they are in fact much more complicated, being composed of a number of different parts with specialized functions. These parts include mitochondria, used in cellular respiration, and (in some) chloro-plasts which convert energy from the sun into chemical energy during the process of photo-synthesis. Chloroplasts occur in green plants and are responsible for their colouring but when they occur in protozoa they may also be yellowish or brownish. Protozoans are ex-tremely active and move about in one of three ways: by means of long, whip-like flagella, through the beating of short filaments called cilia or, as in the amoeba, by extending the cytoplasmic processes of the body and allow-ing the nucleus to flow into them.

Living protozoans are found in soil and fresh and salt water. Some protozoans are solitary but other such as *Volvox* (page 339) live in colonies or aggregates. This habit is more developed in the sponges (Phylum Po-rifera). Sponges are composed of many cells living together in a permanent and organized manner; such animals are known as met-azoans. The structure of these metazoans is primitive for, although the cells adhere to one another to form a definite cup shape with certain cells performing special functions, they lack organs and have no nervous system.

The outer layer or skin of a sponge body is formed by cells and covered with tiny pores. Water passes into the body through the pores and then out through larger openings on the surface called vents. The water is taken in and forced out by means of collared cells, each of which has a mobile whip, or flagellum. It is the action of thousands of these that stimulates the passage of water. As the water passes through the body, tiny food particles and oxygen are filtered out of it by the collared cells and passed on to amoebocytes which digest it and pass it on to other parts of the body. The amoebocytes, as their name im-plies, resemble the amoeba in as much as they are constantly changing shape. These cells wander freely about the body. Besides their food-distributing function they have other duties such as the construction of the skeleton.

The skeleton of a sponge is the structure that is seen from the outside. It may be one of three types. In the first type (calcareous sponges), small spikes called spicules are formed of lime. These are produced within the body of the sponge and carried by amoebocytes to their position. The same process happens in the second type (glass sponges) but in this case the spicules are formed of silica. In the delicate network of Venus' Flower Basket Sponge, the spicules are often star-shaped and combine to form the structure surrounding the cells and supporting the body. In the third type, the skeleton is composed of protein fibre and is soft and flexible; these fibres are produced where they are needed. This last type of skeleton is probably familiar to most people as it is the one found in the species commonly used as the bath sponge.

The external appearance of sponges varies considerably, from simple crust-like structures on stones and corals to tree-like or funnel-like forms, most of which in the adult stage are attached and sedentary. These complex structures were for centuries regarded as plants. Although the Greek philosopher Aristotle had recognized them as animals, it was not until the 1850s that they were finally accepted as such.

Some members of the coelenterates are also often mistaken for plants or even inanimate objects. This phylum, Coelenterata, includes jellyfishes, sea anemones and corals. Some of them are colonial animals but all have a primitive nervous system, muscle fibres and alimentary tract. The group is divided into three classes: the Hydrozoa which includes the Hydra and Portuguese Man o' War; the Scyphozoa (jellyfishes); and the Anthozoa (sea anemones, sea fans and corals).

The coelenterates are polymorphic, that is they exhibit two different forms at different stages of their life: the polyp and the medusa. The polyp is a bag-shaped structure attached by a disc to rocks or other objects. It has one opening at the top, the mouth. This is surrounded by tentacles which capture prey and push it through the mouth into the body cavity. The medusa is generally saucer-shaped and floats mouth-downwards so that its tentacles hang down into the water. In both forms the body wall is formed by two layers of cells between which is a layer of jelly. The tentacles possess special capsules called nematocysts, containing coiled threads which shoot out and pierce prey, paralysing and in some cases killing it before it is ingested.

Some coelenterates, such as jellyfishes, spend most of their life as a medusa. Others, such as the corals and sea anemones (page 338), spend their whole life as a polyp. There are however many species that exhibit a more typical life-cycle with alternation between the two phases. One such species is the colonial hydrozoan *Obelia*, found in salt water. In the polyp stage it is attached by a stem to rocks or algae. From the stem base arise branches which expand at the end into polyps, some of which take in and digest food while others produce medusae. The medusae leave the polyp to swim freely in the water, some producing eggs and others spermatozoa. From the fertilized egg emerges a larva known as the planula which soon attaches itself to a suitable object and develops into a polyp which will form a new colony.

Annelids and Molluscs

The emergence of the true worms, or annelids, was an important step forward in the evolution of the animal body plan which eventually gave rise to the arthropods, one of the most successful groups of invertebrates. Annelids have a segmented body. The segments are self-contained units with similar organs in each, including a pair of excretory organs (nephridia) which remove waste from the body. Some other organs are centralized in certain areas of the body. The segmentation is visible externally in the grooves or rings placed at more or less even intervals along the body. Between the digestive tract and the body wall there is a cavity called the coelom. The coelom is filled with fluid which, among other functions, takes digested food substances from the alimentary tract and distributes them to the body tissues. As well as the alimentary or digestive tract, annelids have a blood vessel and nerve cord running from the head to the hind end of the body, thus integrating the whole structure.

Although true worms are soft-bodied, their skin cells produce a soft cuticle that covers the entire surface.

Annelids are divided into five groups or classes, the most important being bristle-worms, pondworms and earthworms (page 334), and the freshwater and salt water leeches. The bristleworms, class Polychaeta, have limb-like projections surrounded by stiff hair-like bristles. One of these, the Lugworm, lives in sandy or muddy habitats where it burrows into the ground, constructing an L-shaped tunnel. During the digging process the worm ingests both organic material and sand. Waste products are deposited in a heap at the entrance to the burrow. Leeches (class Hirudinea) are blood-sucking worms. They are partially parasitic for, having taken blood from their victim or host they resume an independent existence.

Like the annelids, molluscs are soft-bodied animals despite the presence in most of a hard outer shell. One of the main characteristics of these invertebrates is the presence of a mantle, or skin, covering the body like an envelope. The shell is composed from calcareous substances secreted by the mantle. In most molluscs the shell is an obvious feature. These include the single-shelled limpets, cowries, whelks and snails, and bivalves such as mussels and oysters. In the higher, more developed molluscs, the cephalopods (squids and octopuses), the shell is either vestigial and internal (the squid) or absent (most octopuses). A cavity between the mantle and the body contains the respiratory organs, gills and other structures. The main locomotory organ is the muscular foot which usually protrudes from the ventral surface of the body but in cephalopods has become modified into arms or tentacles. Molluscs form the second largest group of invertebrates after the arthropods and are divided into five classes.

Echinoderms and Arthropods

Echinoderms are entirely marine animals found on the sea-bed, especially in warm and tropical waters. This group contains the starfishes, sea-lilies, sea-cucumbers, brittle-stars and sea-urchins, all of which display radial symmetry in the adult stage. Five arms radiate from a central disc, a feature most obvious in the starfishes (page 328) and brittle-stars but also present in the other members in modified forms. In the sea-urchin, for example, the arms are bent upwards and united so that the body appears spherical. This echinoderm also exhibits another characteristic of the group, the presence of a skeletal plate called a test which is covered with spines (the name echinoderm means spiny-skinned). The test is often seen washed up on the shore, minus its spines. The spines are flexible and attached to the test by muscles; with the tube feet mentioned below, they aid locomotion and help the animal to keep upright.

Except in the sea-lilies (class Crinoidea), the mouth is on the underside of the body, situated in the centre and composed of five parts. On the underside of the limbs are tiny tube feet connected to canals. Tube feet use water pressure to produce suction for movement and are also used for feeding by gathering particles from the water and passing them on to the mouth, and in respiration.

The Phylum Arthropoda is the largest in the animal kingdom. Its members, numbering over 850,000, include the enormous group of insects (pages 289–324), crustaceans such as lobsters, crabs and shrimps, and the arachnids such as spiders and scorpions.

An arthropod's body is segmented. In most, each segment possesses highly specialized organs and, in a typical arthropod, one pair of appendages. Insects, crustaceans and millipedes usually have three distinct segments: the head, thorax and abdomen. In some, the head and thorax may be combined to form a cephalothorax. Insects invariably have three pairs of legs but in other classes the number of limbs varies enormously, even within a group.

Two other important features of arthropods are the external skeleton or exoskeleton, which is composed primarily of chitin and covers both the legs and body, and the flexible connective tissue between the jointed legs and segments which allows movement.

Phylum Echinodermata, including sea urchins, sea cucumbers and sea lilies. Class Asteroidea, the starfishes. Common Starfish (*Asterias rubens*), approximately 5–10cm (2–4in) in diameter, is found in ponds on rocky shores. Diet includes mussels, barnacles and clams. This starfish has an extremely interesting method of feeding, being able to protrude its stomach through its mouth and digest its prey while it feeds. The Starfish opens the shells of bivalves such as mussels by attaching its suction discs to the shell and exerting a tremendous pull. In members of this genus, *Asterias*, the two sexes are distinct. Both male and female possess two reproductive organs on each arm. From these either eggs or sperm are discharged.

STARFISHES

Starfishes form one small part of the phylum Echinodermata which contains more than 5000 species of spiny-skinned invertebrates. Members of the class Asteroidea, they are found in most of the waters of the world on rocky shores or in shallow waters and many of them burrow in sand or gravel or live among rocks. As in most other echinoderms they display five-rayed symmetry, their limbs radiating from a central disc, and possess tube feet composed of hundreds of tiny water-holding organs which provide suction for movement. Like their close relatives the brittle stars, starfishes have an amazing capacity to regenerate lost limbs. If they are deprived of as many as four of their arms they can produce new ones provided that the central disc remains intact and attached. The Common Starfish is found in ponds on rocky shores, spending the day in crevices and foraging for food at night.

SPIDERS

Spiders are often mistaken for insects but may be very easily distinguished from their fellow arthropods by the presence of four pairs of legs and the division of the body into two segments: the prosoma, or cephalothorax, and the abdomen. (Insects have three pairs of legs and a body segmented into three parts.) Several species of spider are found in gardens and houses in the temperate climates of Europe: the Garden or Cross Spider and the domestic spiders. The former constructs an orb-like web with some threads radiating from the centre point and others arranged in a spiral. When an insect becomes trapped on the gummed surface, the spider moves quickly towards it and wraps it in a silken cocoon before removing it to its retreat where the prey is devoured. The spiders found in baths are usually of the genus *Tegenaria*. Most often these are in search of water or are males in search of a female. These spiders construct a sheet web.

Phylum Arthropoda. Class Arachnidae. Order Araneae. Family Araneidae. Spiders lack antennae but in their place possess a pair of limb-like structures called pedipalps which in the male also function as transmitters of sperm. Prior to mating, the male deposits sperm on the web which he then sucks up with the pedipalps. During mating he inserts this into the female's genital opening. Also located on the cephalothorax are the mouth parts and legs. The abdomen bears projections known as spinnerets through which strands of silk are drawn from the silk-producing glands. Garden Spider (*Araneus diadematus*); female lays 400–800 eggs in crevices. The eggs hatch in spring.

CENTIPEDES

The elongated, flattened body of a centipede is clearly segmented. Each segment, except the two last ones and the one behind the head, bears one pair of jointed walking legs. In most species the legs of the first segment have been modified into fangs or claws called maxillipeds. These emit a toxic substance, an adaptation to a predacious life. All centipedes are carnivorous and prey on invertebrates and other small animals. The largest species are found in the tropics, one South American centipede reaching a length of some 26 centimetres (10.5 inches). Common species are those of the genera *Lithobius*, *Scutigera* and *Geophilus*. Like other centipedes, they move rapidly and may be found under stones, leaves and logs.

Phylum Arthropoda. Class Chilopoda, centipedes. (Sometimes placed with the millipedes in Myriapoda; otherwise the millipedes are classed as Diplopoda). House Centipede (*Scutigera forceps*), about 5 cm (2 in) long, has a brownish body with three dark stripes and 15 pairs of legs. The number of segments, and therefore legs, varies from one species to the next and ranges from 8 segments to more than 100. *Scolopendra gigantea* is found in Central America; it feeds on large insects, lizards and small mammals. The centipede's bite is often painful but never fatal to man. Millipedes are distinguished by having two pairs of legs on almost every segment. Their bodies are generally more cylindrical and almost all of them are exclusively vegetarian.

MANTIS SHRIMP

This marine animal belongs to the group of arthropods known as crustaceans and is therefore related to the crabs and lobsters. The elongated body has a long abdomen and seven segments in the thorax. The second pair of thoracic limbs are larger than the others and the terminal segment has a sharp point; the mantis shrimps take their name from the resemblance of these limbs to the front legs of mantids. At the end of the abdomen there is a tail-like fan and the second pair of antennae have stalked eyes. Mantis shrimps are most abundant in tropical waters but a few species are found in the Mediterranean Sea and one occurs off the coast of Cornwall, in Britain. They are bottom-dwellers that burrow in the sand.

Phylum Arthropoda. Class Crustacea. Order Stomatopoda, the mantis shrimps. These crustaceans are voracious eaters. They feed on smaller crustaceans which they capture with their large, sharp thoracic limbs. Extremely aggressive, they can give a painful wound to man and for this reason are called 'Split-thumbs' in the West Indies. *Squilla mantis* and *S. desmaresta* are found in the Mediterranean; the latter also off the coast of Cornwall. *Pseudosquilla biglowi* occurs off the coast of Southern California.

HERMIT CRAB

True crabs and other crustaceans are characterized by an external skeleton that completely covers the body and legs, but in hermit crabs the hard cuticle does not extend to the soft abdomen so that it is left unprotected. To render itself less vulnerable the crab takes on the abandoned shells of gastropod molluscs (see page 336), especially those of whelks and winkles. It grasps hold of the shell with its tail pincers and pulls its body inside, the head and limbs projecting from the opening so that it is free to move about. Most hermit crabs form mutually beneficial relationships with sea anemones and sometimes other animals. The Common Hermit Crab places a species of anemone on top of the shell, the stinging tentacles of the anemone affording it further protection from predators. The anemone benefits by living off scraps of food left by the crab.

Phylum Arthropoda. Class Crustacea. Order Decapoda. Family Paguridae. Common Hermit Crab (*Pagurus bernhardus*) has a body length of up to 10cm (4in). Like most other crustaceans it is marine and is found on the shore. In hermit crabs the body is elongated, with a tail fan on the hind part of the abdomen. There are five pairs of legs. The first pair are modified to form pincers; the last two pairs are short and are not used for locomotion. It also forms a symbiotic relationship with the ragworm *Nereis fucata* which lives inside the shell and feeds on small organisms on the crab's body.

CLEANER SHRIMP

Phylum Arthropoda. Order Decapoda. Suborder Natantia, comprising shrimps and prawns. Shrimps are generally found in sandy waters where they burrow in the sand during the day, venturing out at night to feed on worms, mollusc eggs and other small animals in addition to scavenging. The abdominal limbs are well-developed for swimming and most shrimps possess only one pair of pincers. Members of the family Stenopodidea, to which *Spongicola venusta* belongs, have 3 pairs. European species include the Common Shrimp (*Crangon vulgaris*), Sand Shrimp (*C. crangon*) and prawns of the genus *Palaemon*.

Shrimps, like crabs, are scavengers that help to dispose of debris on the sea floor, although most also take live prey. The species known as cleaner shrimps have become totally adapted to feeding on parasites of fishes and some other sea-dwelling animals. Many of them are brightly coloured to attract the attention of fishes which swim up to them and allow the shrimps to move over their bodies and sometimes into their gills. *Stenopus hispidus* is the largest species. It has a brightly coloured red and white striped body, with typically long antennae and antennules which it uses to signal to fishes. Another species of tropical waters is *Spongicola venusta*, the larva of which ventures into the body cavity of the Venus Flower Basket Sponge. It remains there for the rest of its life, its size prohibiting it from leaving.

333

EARTHWORM

Earthworms belong to the phylum Annelida, the segmented worms, those of the family Lumbricidae being the common species found in almost every part of the northern hemisphere. Because of their habit of burrowing in the earth, they are welcomed by gardeners for whom they provide a valuable service by mixing and aerating the soil. The earthworm has no respiratory organs and breathes by taking in gases through the skin which is covered with a mucous substance. Similarly, the sense organs occur all over the body in the form of specialized cells attached to sensory hairs in the skin. The earthworm feeds on vegetable and animal matter, emerging from its burrow at night to forage in the area near the opening. It is well-known for its ability to regenerate lost parts.

Phylum Annelida. Order Oligochaeta. Common Earthworm (*Lumbricus terrestris*). Like other earthworms this species is hermaphroditic, each individual possessing both male and female reproductive glands, but it is not self-fertilizing. During the breeding season certain segments become swollen and produce a mucous sac. Two worms encountering each other in the same condition will lie side by side. Their anterior sections become enclosed in a mucous tube, enabling them to exchange sperm. The earthworms then begin to wriggle out of their respective sacs. As they do so, the eggs pass out of their bodies into the sac and are fertilized by the sperm. When the sac is detached it forms a cocoon in which the embryos develop.

OCTOPUS

Phylum Mollusca. Class Cephalopoda, squids and octopuses. Order Octopoda contains the octopuses; Order Decapoda the squids and cuttlefishes. *Octopus vulgaris* is found mainly off the coasts of Africa and Central America but sometimes occurs off the coast of southern England. Its diet includes crabs and lobsters which it grasps in its tentacles or arms as it feeds. It inhabits rocky crevices on the sea-bed and lays its eggs in capsules which hang from rocks. The larvae are free-swimming for several weeks. North American species include *O. apollyon* which occurs off coasts in the Pacific Ocean. It may weigh up to 45.4kg (100 lb).

The Class Cephalopoda comprises the squids, octopuses and cuttlefishes; they are closely related to bivalves (page 336). Although they appear markedly different, the body is basically similar but more highly developed. In cephalopods, the foot found in the lower molluscs has evolved into tentacles around the head and the mantle into a bag-shaped structure behind the head. This has enabled them to become far more active and also aggressive predators. Octopuses live on the rocky sea-beds of warm and tropical waters, although at least one species occurs off the coast of southern England. It moves about either by crawling over the bottom or, as in other cephalopods, by producing a jet of water from the mantle cavity which propels it forwards or backwards, according to the direction in which the jet flows. Like the squid, the octopus is able to emit a pigmented fluid which conceals it from predators.

BIVALVES

Bivalves include cockles, mussels and oysters, all of which produce two shells, or valves, from secretions of the mantle which envelops the body. These valves are termed 'right' and 'left', not 'top' and 'bottom', for they are hinged along the dorsal surface. The ringed ridges on the shell indicate stages of growth, the smallest ones being the earliest formed; as the mollusc grows it adds further rings. Bivalves do not have a head; they feed by filtering particles from the water in the mantle cavity. The water also provides oxygen which is taken up by the gills. The mantle edges, or skirt, contains organs sensitive to touch and taste. In some molluscs, such as the scallop, the skirt is also equipped with many eyes. Mussels attach themselves to rocks and other hard surfaces by means of a thin thread-like structure called the byssus. Others, such as carpet shells and cockles, bury themselves in the sand.

Phylum Mollusca. Class Bivalvia (Pelecypoda). Superfamily Mytilacea, the mussels. Superfamily Pectinacea, scallops and file shells. The scallops' brightly coloured shells are not symmetrical, one being domed, the other flattened. More active than most, it swims by opening and shutting the valves and propelling water from the interior in a jet. Superfamily Ostreacea, the oysters. The family Osteridae contains the edible species, which are of great commercial importance and are found in both cultivated and natural beds. Pearls of commercial value are produced by species in tropical waters, although other species may give rise to insignificant specimens.

SNAILS

Snails are closely related to slugs and together they form the class Gastropoda. Snails possess a single shell but in slugs this feature is lacking. As in most other molluscs, locomotion is provided by a foot. In snails and slugs locomotion is made easier by the secretion of slime from a gland situated under the mouth, producing the characteristic silvery trails on the ground. Situated on the head of the snail are two pairs of tentacles or horns. The hind pair contain 'eyes' at the tip. These are sensitive to light, but cannot really be called sight organs. Many snails, such as the Edible or Roman Snail illustrated here, can tolerate dry and cold conditions. Others require damp habitats; these are found on the banks of streams and on marshland.

Phylum Mollusca. Class Gastropoda comprising snails, slugs and limpets. Edible or Roman Snail (*Helix pomatia*) occurs on chalk downs and usually hibernates in the winter, closing off the opening to the shell with a mucous substance. Like most other land snails it breathes through a lung in the mantle cavity. The Garden Snail (*H. aspersa*) is considered a pest by gardeners as it eats the leaves of cultivated plants. Like the former species, it is sometimes eaten by humans. Species that prefer moist conditions include the Amber Snail (*Succinea putris*) which is found in marshlands and other damp habitats.

SEA ANEMONE

Because of their plant-like appearance these attractive animals were given the name of a flower. They belong to the phylum Coelenterata and are closely related to jellyfishes and corals. Most of their life is spent attached to rocks and other hard surfaces although they are able to creep along on the basal disc. The body is a cylindrical column with a hollow centre and slit-like mouth at the top around which are arranged the hollow tentacles. As in other members of the group, the tentacles contain nematocysts, cells which paralyse prey; the tentacles then convey the food to the mouth. Sea anemones are most abundant in warm waters but some occur in temperate seas; in the waters off Great Britain there are over 30 species.

Phylum Coelenterata. Class Anthozoa includes the sea anemones and corals. Order Actiniaria. Plumose Anemone (*Metridium senile*) is about 20cm (8in) high. Colour is orange or white and the tentacles are feathery. Members of this genus are found in Europe and North America. Two genera, *Adamsia* and *Calliactus*, live on the shells inhabited by the Hermit Crab. Sea anemones may reproduce in one of two ways: in some the young develop inside the parent and are then released into the water; in others buds develop which break off to live an independent existence.

Phylum Protozoa, of which five examples are shown below. *Euglena* is a marine plant-like protozoan which in large masses gives a greenish tinge to water. The cell contains chromatophores of chlorophyll and manufactures its own food by photosynthesis. The obvious eye-spot, or stigma, is light-sensitive. Movement is provided by a single flagellum. *Chlamydomonas* occurs in freshwater and, like *Euglena*, contains chromatophores. It has two flagella. In *Amoeba*, *Paramecium* and *Euglena* the cytoplasm is divided into two parts, the outer ectoplasm and the central mass, or endoplasm, which is more fluid. *Paramecium* feeds mainly on bacteria and moves by beating the cilia on the surface which also waft food particles into the gullet. *Amoeba* moves by sending out finger-like protrusions of the cytoplasm which the nucleus moves into. It feeds by flowing around its prey and then enclosing it in a tiny vacuole. In *Volvox* several hundred individuals form a wall enclosing a jelly-filled interior.

PROTOZOA

The phylum Protozoa contains some 46,000 named species of single-celled organisms, many of which have highly complex structures with specialized parts known as organelles. Some live as solitary units while others, such as *Volvox*, form communities. Because of the plant-like characteristics of many of these unicellular organisms, some authorities place them in their own kingdom, the Protista. Protozoans are found in soil, freshwater and salt water; some live as parasites in other organisms. Each species appears to favour a particular kind of habitat. They occur all over the world wherever there is an adequate supply of moisture, although some species of Sporozoa (the spore-formers) are able to tolerate drier conditions. The parasitic species are known to exist in every higher animal including man, particularly in the digestive tract and the blood where they may cause diseases such as sleeping sickness and malaria.

1. *Euglena*
2. *Chlamydomonas*
3. *Volvox*
4. *Paramecium*
5. *Amoeba*

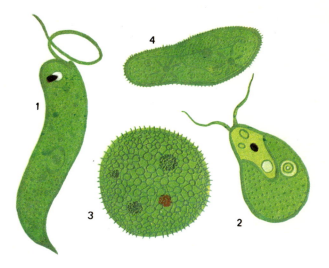

INDEX

Acknowledgements

Anthonissen C.: 173, 133; Ardea/Beames I.R.: 68; Ardea/Taylor R. & V.: 264; Arnhem R. en D.: 17, 30, 60, 114, 116, 117, 118, 119, 120, 122, 128, 130, 131, 134, 136, 141, 142, 143, 144, 145, 146, 147, 151, 153, 154, 155, 156, 157, 162, 163, 166, 168, 169, 172, 177, 178, 181, 184, 185, 186, 187, 188, 190, 191, 192, 193, 195, 199, 200, 201, 202, 205, 208, 209, 212, 214, 215, 216, 217, 218, 219, 212, 234, 250, 289, 292, 293, 297, 298, 299, 306, 307, 321, 372; Bijnens A.: 13, 231, 242, 266, 267, 268, 270, 271, 272, 273, 274, 275, 276, 279, 280, 281, 282, 283, 284, 286, 287, 288, 328, 331, 333; Bonhivers G.: 51, 57, 77, 80, 96, 98, 99, 106, 115, 125, 139, 295; Bresseleers: 197, 198, 204, 222, 329; Cauwels M.: 174, 196, 207, 183; De Vocht E.: 16, 23, 25, 34, 44, 45, 46, 48, 49, 55, 58, 64, 71, 73, 74, 75, 76, 79, 85, 91, 93, 94, 95, 103, 104, 105, 132, 138, 148, 149, 150, 152, 232, 236, 243, 305, 332, 338, 339; De Wavrin H.: 15, 21, 22, 32, 43, 65, 210, 227, 247, 254, 259, 262, 301, 330; Gilis L.: 126, 159, 160, 203, 311, 319; Goris F.: 50, 127, 129; Hautekiet M.: 161; Hoyaux R.: 84, 89, 337; Lehaen P.: 180; Luyckx B.: 133, 135, 137, 158, 211; Mazza G.: 26; Meli-Park: 170; Piron P.: 47, 52, 63, 70, 107, 179; Press & Pictures (P&P)/Ausloos H.: 37, 56, 100, 102; P&P/Cordier J.: 24, 41, 318; P&P/Danegger M.: 78; P&P/Duscher E.: 62, 67, 86; P&P/Garot C.: 18, 54, 233, 304; P&P/Lemoine L.: 38, P&P/Tastenay J.: 35; P&P/Van der Vaeren J.: 53, 59, 265, 277; Raes F.: 69, 241, 263, 82, 83, 92; Samyn A.: 228, 229, 230, 235, 240, 244, 249, 251, 252, 253, 255, 256, 257, 258, 323; Tessloff: 336; Van Autenboer T.: 101; Van Elsaeker L.: 164; Van Rooy L.: 337; Vorsselmans F.: 312, 313, 314, 315, 316, 317, 326, 290; Vorsselmans L.-Verbruggen: 19, 20, 33, 36, 121, 123, 291, 296, 302, 303, 308, 309, 320, 336; ANTWERP Zoo: 27, 28, 237.